BEE GARDEN

Getting up close to bees

My love of bees was kindled by admiration for the inconspicuous solitary mason bees rather than the highly efficient, colonial honey producers. The place I discovered them was a real surprise. I found them in the balcony railings of my student flat, where they were using little holes for their nests. The balcony was actually a loggia, which sounds like something from the sunny south but bore about as much resemblance to this as an indoor swimming pool does to the Mediterranean. It was dark and sad and the sun was a rare and honoured guest, yet there was life there. My interest in bees had been aroused and I expanded my small, determined colony by providing horizontal bundles of bamboo canes. Later a whole host of tree bumblebees took up residence in the wall of the building, and I shared the balcony with them for a whole summer, as if they were part of my family. It seemed that very little space is required in order to help these creatures. The aim of this book is to show you how to make your garden attractive to bees. It is not intended to be a reference book, but it presents the commoner species and describes how you can encourage them. It also gives the best planting schemes. But watch out! Getting involved with these furry insects can be addictive. Elke Schwarzer

© Elke Schwarzer 2020
Elke Schwarzer has asserted her right to be identified as the author of this work.

Translation from German: Rae Walter in association with First Edition Translations Ltd, Cambridge, UK

All rights reserved. No part of this publication may be reproduced or stored in a retrieval system or transmitted, in any form or by any means, electronic, mechanical, photocopying, recording or otherwise, without prior permission in writing from Haynes Publishing.

First published in the UK in February 2020

British Library Cataloguing in Publication Data
A catalogue record for this book is available from the British Library.

ISBN 978 1 78521 696 1

Library of Congress catalog card no. 2019944197

Published by Haynes Publishing,
Sparkford, Yeovil, Somerset BA22 7JJ, UK
Tel: 01963 440635/Int. tel: +44 1963 440635
Website: www.haynes.com

Haynes North America Inc.
859 Lawrence Drive, Newbury Park, California 91320, USA

Printed and bound in Malaysia

While every effort is taken to ensure the accuracy of the information given in this book, no liability can be accepted by the author or publishers for any loss, damage or injury caused by errors in, or omissions from the information given.

BEE GARDEN

ALL YOU NEED TO KNOW IN ONE CONCISE MANUAL

Elke Schwarzer

Contents

A garden for bees	6
The garden as a refuge	8
Flowers for a busy garden	9
Designing a bee garden	11
Bee-friendly gardening	16
Wild bees	18
Tawny mining bee	20
Alpine currant	21
Wool carder bee	22
Motherwort	24
Downy woundwort	25
Betony	26
Lamb's ear	27
Hairy-footed flower bee	28
Creeping comfrey	30
Fumewort and holewort	31
Lungwort	32
Early-flowering borage	33
Davies' colletes	34
Achillea filipendulina	35
Yellow chamomile	36
Tansy	37
Ivy bee	38
Common ivy	39
Sweat bees	40
Yellow-faced bee	42
Leaf-cutter bees	44
Bladder-senna	45
False indigo	46
Broad-leaved everlasting pea	47
Red mason bee and European orchard bee	48
Solitary mason bee	50
Clustered bellflower	51
Trailing bellflower	52
Giant bellflower	53
Large-headed resin bee	54
Elecampane	55
Violet carpenter bee	56
Long-leaved bear's breeches	57
Chinese wisteria	58
Clary sage	59
Bumblebees	60
Common carder bee	62
Tree bumblebee	63
Buff-tailed bumblebee	64
Garden bumblebee	65
Aquilegia	66
Korean mint	67
Russian sage	68
Turkish sage	69
Monk's hood	70
Common foxglove	71
Perennial cornflower	72
Honeysuckle	73
Obedient plant	74
Bergamot	75
Bastard balm	76
Faassen's catmint	77
Viper's bugloss	78
Balm-leaved archangel	79
Common bugloss	80
Japanese rose	81
Common sage	82
Welsh poppy	83
Cup plant	84
Purple gromwell	85
Hotspot plants	86
Apple	88
Marjoram	89
Culver's root	90
Alder buckthorn	91
Ice plant and orpine	92
Hardy fuchsia	93
Wallflower	94
Wild teasel	95
Mountain fleece	96
Common poppy	97
Globe thistle	98
Lavender	99
Purple tansy	100
Scots rose	101
Many-flowered rose	102
Bee tree	103
Hollyhock	104
Cranesbill Rozanne	105
Hemp agrimony	106
Almond willow	107
Filler plants	108
Wild garlic	110
Leopard's bane	111
Ground ivy	112
Bluebells	113
Ornamental onion	114
Spring snowflake	115
Agrimony	116
Snowdrop	117
Greater celandine	118
Knotted cranesbill	119
Wood forget-me-not	120
Hedge woundwort	121
Index	125

A garden for bees

Honey bees, bumblebees or wild bees – they all love flower gardens. Many wild bees will even come to stay if they can find suitable nesting places in the garden, as well as their favourite flowers.

The garden as a refuge

Life is no picnic for bees any more. Honey bees are under attack from the *Varroa* mite, which has become Public Enemy No. 1 and is causing great damage. What's more, it is not only the *Varroa* mite that threatens bees; they are also susceptible to modern agricultural pesticides.

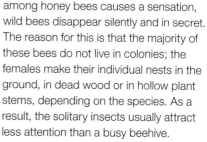

These pesticides also affect bumblebees, because they fly long distances and thus come across poisoned fields. Flowery landscapes are rare; the countryside is dominated by cornfields and fertilised, flowerless meadows. While the death toll among honey bees causes a sensation, wild bees disappear silently and in secret. The reason for this is that the majority of these bees do not live in colonies; the females make their individual nests in the ground, in dead wood or in hollow plant stems, depending on the species. As a result, the solitary insects usually attract less attention than a busy beehive.

The solution: invite the honey bees' wild relatives into your garden. Of course, you will not be able to harvest any honey from them, but in return there will be no need for you to take a beekeeping course and you will meet a wider range of flamboyant bee personalities. This is because once the wild bees have arrived, they will bring many other species in their wake, including brood parasites that make life easy for themselves, such as the ruby-tailed wasp with its metallic sheen.

Luring bees to your garden will provide hours of entertainment. You will get to see fighting drones, acrobatic displays near the nesting aids, the gold-fringed mason bee juggling with the snail shells in which it nests, and many skilled worker bees that banish all prejudices concerning the competence of the female sex.

To top it all off, these insects are experts at pollinating garden plants, often making an even better job of it than honey bees. So read on, and find out how you can transform your garden into a colourful refuge for many types of bee.

Flowers for a busy garden

Many wild bees are quite particular and have favourite plants. Bumblebees and honey bees, by contrast, are more flexible and make use of many different flowers throughout the year. Here, you can read about the best way to feed all the visitors buzzing around in your garden.

The right plants

Bees need single flowers or those with few petals in order to feed effectively, and that often means natural varieties. The differences between natural and cultivated varieties can be seen in roses. Wild roses have just five petals and many anthers laden with pollen. By contrast, cultivated varieties often have so many petals that they look like pom-poms. That said, cultivated roses have one advantage: whereas wild roses usually rest after one brilliant flowering and quietly work on forming their fruits, repeat-blooming roses flower throughout the summer.

What's more, there are many plants that bees avoid altogether, such as forsythia and mock orange. So, it is important to plant the correct species.

How important are native plants?

Many wild bees are adapted to specific species of flora and prefer to feed locally. When selecting plants, a crucial factor is whether foreign plants have close relatives in the native flora. Plants from the Mediterranean area are often a good choice. For example, one of the mason bees (*Osmia adunca*) visits the Mediterranean purple viper's bugloss (*Echium plantagineum*) as well as the common viper's bugloss (*E. vulgare*), while the yellow-faced bee (*Hylaeus species*) likes not only the native tansy, but also the cultivated achillea that originates from the Caucasus – both belong to the same subfamily within the Asteraceae. Specialists such as the solitary mason bee (*Osmia rapunculi*) and the gold-tailed melitta (*Melitta haemorrhoidalis*), meanwhile, delight in bellflowers of all kinds. In fact, you can't go wrong with bellflowers, or with legumes.

Another advantage of providing a range of plants from around the world is that it ensures continuous blossoming from high summer onwards, even after the native plants have finished flowering.

How many plants?

As a rule of thumb: the more specialised a wild bee is, the more specimens you need to plant. However, even the generalists among bees are grateful when they find many of the same flowers together, so they don't have to learn a new way to reach the nectar for each species.

In small gardens, you can cheat by choosing long-flowering species, such as perennial cornflower, larkspur, meadow cranesbill, catmint and sage, or cutting the plants hard back after the first flowering to make them flower a second time and ease the supply situation.

Flowers for bumblebees

Bumblebees live in colonies and need flowers from March to October. As soon as the lime trees blossom in June, their numbers become so great that they can

↑ Caught in the act: a buff-tailed bumblebee breaking into an aquilegia.

no longer find enough food – the summer gap has arrived. At that point there will be a lot of dead bumblebees lying under the trees, so include plants that continue blooming in summer in your plans. The native bumblebee species in the garden are not choosy. They can't afford to cherry pick, either, because what's on offer is continually changing. Nor do they compete with each other, since the various species of bumblebee have proboscises of different lengths.

Typical bumblebee flowers – such as sage, dead-nettle, aquilegia, foxglove, antirrhinum and monkshood – have their food prepared for bumblebees with proboscises of a particular length. As well as planting these, try to provide open flowers for species with short proboscises – for example willows, roses and *Asteraceae* such as echinaceas and silphiums. Don't worry too much, though; insects that are challenged in terms of the length of their proboscises simply gnaw an opening in the flower from the outside. You can see evidence of these break-ins particularly well on the aquilegia, corydalis and comfrey. As a bonus, there is a whole series of followers – such as small wild bees – who are looking for a free ride and are also make use of the perforations.

Roses, lime trees, cranesbill, poppies and hollyhocks offer a lot of pollen and are profitable plants for bumblebees, especially if they employ buzz pollination. This is when bees gather a bundle of stamens together and shake them at high frequency, making the pollen fall right into their laps.

Designing a bee garden

Wild bees are appreciative visitors to the garden and, being busy bees, they like to make their nests close to their food-plants. This means that you can provide them with board and lodging even in the smallest garden.

The good news about planting for bees is that you don't need to worry about gaps or 'weeds' in your lawn; bees are not interested in bowling greens, so don't become a slave to lawn care. In fact, ground-dwelling bees like patches of thin grass. The same is true of some flowers; crocuses feel better when the dominant grasses don't come too close to their corms. They can be planted in autumn immediately below the surface of the lawn and will cater for the first bumblebees of the season, as well as other bees. Later on, white clover is the winner. Even the daisy is a valuable supplementary food for male wild bees and sweat bees.

On unfertilised lawns that do not have to look perfect you can allow the presence of ribwort plantain and hawkweed, especially orange hawkweed (*Hieracium aurantiacum*). Since the grasses grow more slowly because the soil is lower in nutrients, there is no need for weekly mowing, which allows the plants to come into flower.

Even better are flower meadows on poor soil that feature bellflowers, meadow-sage, devil's-bit scabious and field scabious. By sowing yellow rattle (*Rhinanthus minor*), which is hemiparasitic on grass roots, you can keep the dominant grasses small. Do not mow until September, when most wild bees have finished their family planning and the plants have seeded. Remove the cut grass so that the soil does not receive any fresh nutrients. If the soil has not yet become poor enough, the meadow buttercup will certainly appear, providing food for *Osmia florisomnis*, another of the mason bees.

Doubly useful: fruit and flowers

Whatever your favourite fruit or vegetable, bees see to it that your table is well stocked. And even in small quantities, home-grown fruit and veg that have not been sprayed help bees that arrive from outside the garden. You can even see mishaps in a positive light: if cabbages, basil or lettuce run to seed, at least there is something for the bees! In fact, allow a few onions and chives to flower to provide pollen for the onion yellow-faced bee (*Hylaeus punctulatissimus*). A herb

⬇ The orange-tailed mining bee (*Andrena haemorrhoa*) likes to nest in threadbare lawns.

spiral not only caters for bees and the kitchen but also offers nesting places for many other species.

Hard facts? Paths and walls

Where ground-dwelling bees are concerned, hard surfaces are bad news since they are often sealed and have poor vegetation cover. So, try to lay natural stone paving and leave large, sand-filled gaps where vegetation can take root. Driveways can be designed to look natural and romantically wild with a gravel surface that can be driven over but has space at the edges for a distinguished community of viper's bugloss, mullein and other plants with a pioneering spirit. Some wild bees love to settle in dry stone walls, which also create islands of warmth in the garden. Small flowering plants can be grown in the crevices. Use limestone for the walls as it acts on plants like a fertiliser.

Home for bumblebees

From March onwards the first queens are on the hunt for a suitable spot for nest building. Cosy hollow spaces such as bird boxes or mouse nests are often popular choices. Some species even nest in tufts of grass or in the exterior insulation of the house. Bumblebees in the house wall cause no damage and usually behave peacefully, so are not a problem.

You can also buy a bumblebee box ready made or, preferably, build a nesting aid yourself, as decribed in the following passages. It is important to control access to the entrance hole with a kind of cat flap for bumblebees. This will keep out wax moths, whose larvae destroy the wax of the brood and honey cells, whereas the bumblebees will quickly get used to the flap and navigate it well.

Bursting with life: dead wood

Old trunks of fruit trees, willow, poplar or birch make perfect Noah's arks. You can either place parts of the trunk directly in a flowerbed or just leave dead trees standing. Being open to attack and not weatherproof is not a defect in dead wood. Beetles will get to work on it first and, later, many different wild bees will use the old tunnels. If the wood eventually feels completely rotten, carpenter bees, fork-tailed flower bees and leaf-cutter bees will find their way in, digging their own tunnels.

Apartment blocks: nesting aids

Wild bees that nest in hollow stems or beetle tunnels are especially easy to help. You can tie reed mats, bamboo canes or other hollow stems in bundles and use wood glue to fasten them in place horizontally inside wooden boxes or tin cans. You don't have to go straight out and plant a bamboo in your garden, either; many shrubs, such as common meadow rue (*Thalictrum flavum*), ornamental onions and common teasel, have hollow stems and provide nesting material. In spring you can cut off the long tubes and trim them to a suitable length to make a nesting aid. A waterproof cover will keep them dry.

You can also drill holes 5–10 cm (2–4 in) deep and 2–9 mm (0.08–0.35 in) wide in well-seasoned blocks of beech, birch or oak. The holes should not be drilled too close together and should be finished cleanly, with no splinters. Do not drill into the end grain as this often results in

→ **Not at all grey and monotonous: dry stone wall with houseleeks, bellflowers and Mexican fleabane (*Erigeron karvinskianus*).**

splitting, but always from the bark towards the centre of the trunk. The holes must not go all the way through the wood.

Airbricks only become attractive to bees if the spaces in them are filled with bamboo stems or light soil. Thin extruded pantiles are suitable for immediate occupation. There is no point in collecting empty snail shells to fill a compartment in the nesting aid. Mason bees that make their homes in these look for them on the ground, turn them this way and that, and camouflage them with plant material, which is not possible in the narrow space of the nesting aid. Bundling elder twigs together and arranging them horizontally is not successful either, as wild bees that specialise in pithy stems look for single vertical stalks in the countryside. It is better to tie individual stems of bramble or mullein to a fence post.

DIY stores have noted the trend and offer 'insect hotels'. These ready-made homes look nice but often turn out to be unfit for purpose. If the drill holes are full of splinters or bored in laminated wood, or if the bamboo stems are not actually hollow, nothing will rescue them. If you make one yourself, however, you will have complete control and you can be sure that the bees will move in.

A bustling beehive can be very interesting for children, which can make some people anxious. However, the buzzy hurly-burly of the bees is not dangerous; most wild bees are so small that their stings leave no lasting impression on human skin. What's more, because the females of solitary bee species are completely on their own and have no big sisters they can call on for help, they will do their best to avoid conflict with us. This means that having bees right next to seating doesn't make that a danger area.

So, build imaginative nesting aids together with your children and plant suitable feed-plants nearby, where you will be able to find 'your' bees. It is good fun and could be the start of a wonderful friendship with bees.

⬇ Aircraft carrier: honey bees filling up with water on a water lily leaf.

Water in the garden

Bees are not amphibious. However, in hot weather honey bees fill up with water to cool the hive, causing their abdomens to swell impressively. This means that a pond is welcome, however small it may be. The water lily beetle, which gnaws holes in the leaves of its food-plant until they look like sieves, is a greedy ally for bees; the more perforations there are in a leaf, the more easily bees can find their own private watering hole.

→ In this summerhouse for bees some of the holes were drilled in the end grain of the wood, which can lead to cracks. However, the airbrick is filled in exemplary fashion.

Bee-friendly gardening

There's no point attracting bees to your garden if in doing so you kill or shun other creatures – it goes against the eco values bee-lovers hold dear. Instead, turn a blind eye to aphids and other pests, and make life easier for bees and yourself.

Pests

The bane of gardeners' lives, aphids are considered a pest by many. Bees, by contrast, love aphids and collect honeydew from them on a grand scale; this is how fir honey is produced, for example. Other beneficial insects and many types of birds are big fans, too, and soon devour the problem if given the chance. If this natural means of aphid control proves insufficient, a fierce jet of water sprayed on the affected plant is always more eco friendly than insecticide.

Because ants like to keep aphids to milk and their underground activity is responsible for loose plant roots, they too are considered villains. However, their activities are actually advantageous for gardeners, since they disperse plant seeds. Ants don't do this just for the sake of it, but because of the delicious food body attached to the seeds. Try spreading crocus seeds near an ant nest – you will be amazed by how quickly the insects make off with them.

Slugs, however, are a major challenge to the pacifist attitude in the garden. Often, they eat only the flowers and leave nothing for the bees. Early-blooming flowers in particular are vulnerable; it is annoying to find all the scillas and narcissi standing sad and flowerless in the bed in the morning. Ferric phosphate slug pellets do not harm hedgehogs or other animals and are the better chemical solution. However, because pulmonate snails also live in my garden and their empty shells are interesting for mason bees (*Osmia*), I avoid pellets and instead make evening attacks on slugs, cutting them up with secateurs. Revolting, yes, but it's selective and non-poisonous.

Rodents – cute but irritating

Bumblebees like to make use of the nests of small rodents for building their own nests. Bank voles are best as their nests

⬇ The lizard thinks the cracks in the wood are great; the bees are not so keen.

↑ Allium sphaerocephalon have a double benefit in the garden: they attract bees but also repel aphids, because of their onion-like scent.

have open exits. That said, mice and voles are not beneficial for bees – I have already lost a few good bee meadows to them – but they have just as much right to the garden as bees so you just have to leave them be.

Is everything a weed?

When it comes to weeds, some species are not as bad as their reputation. Bees apparently think so too. The lesser celandine, for example, is often seen as a yellow peril because it is constantly spreading, but it is a good first source of nourishment for small wild bees that have just come out of hibernation. From May onwards, these tuberous plants die down anyway and are not seen again until the following March. All kinds of dead-nettle are valuable food for bumblebees and are not a nuisance under shrubs. Many wild cranesbills are so tiny as to be almost unnoticeable, but the very smallest species of bee are certain to find them. Only herb-robert, sometimes known as the stinking cranesbill, is too dominant and has to be rooted out in good time.

Throwaway mentality

When replacing rotten pergolas, wooden furniture or posts, look out for bees living in them and don't dispose of them along with the bulky waste. If possible, integrate the wood in the garden. Leaf-cutter bees that tunnel into soft wood are not responsible for the dismal state of garden furnishings; they simply exploit the preliminary work done by fungi.

Willoughby's leaf-cutter bee (*Megachile willughbiella*) often builds its nest in the compacted soil of window boxes and flowerpots. When replanting, don't throw away the old soil – put it in an unobtrusive place and allow the bees to hatch in summer.

Wild bees
Caring loners

Almost all the bees shown here live in the garden of my terraced house. They may not be particularly rare, but they are easy to recognise and observe – and can easily become the gateway drug that leads you into addiction to the fascinating world of wild bees.

Tawny mining bee *Andrena fulva*

In built-up areas mining bees generally have a hard time finding open, sunny ground for their tunnelling activities. However, I have actually seen the tawny mining bee digging between the shrubs in my garden in relatively shady spots. It is also happy to nest in scruffy lawns. Unlike many other mining bees, this species – with its smart red fur coat – is easy to recognise and cannot be confused with others. On the other hand, the slightly built males look very similar to other mining bee drones and can only be correctly classified for certain at mating time, when they have found a female.

In contrast to the striking and beautiful bee, many of its favourite food-plants are inconspicuous: it likes to visit the flowers of currants, gooseberries and even box.

Favourite plants
- Alpine currant (*Ribes alpinum*)
- Barberry (*Berberis*)
- Bilberry (*Vaccinium myrtillus*)
- Cherry plum (*Prunus cerasifera*)
- Common box (*Buxus sempervirens*)
- Gooseberry (*Ribes uva-crispa*)
- Redcurrant (*Ribes rubrum*)
- Willow (*Salix*)

Support
These mining bees are very easy to encourage with plants that are also useful for us and yield a rich crop of fruit, in which the pollinating activities of the bees play a not-inconsiderable part. A soft-fruit garden is guaranteed to attract them. In addition, if you stop turning over the soil the nests will be preserved.

⬇ The tawny mining bee.

BEE PROFILE
- Male 9–12 mm (0.35–0.47 in), female 12–14 mm (0.47–0.55 in) long
- Upper side a striking tawny, underside black
- Pollen transported on the hind legs
- Flying period March to May in one generation
- Nests in tunnels in the ground, which it excavates itself and which may reach a depth of 50 cm (20 in)

Alpine currant *Ribes alpinum*

If you are not too keen on eating currants, just plant the wild form of Alpine currant for the tawny mining bee. She will not consider it a crime if you leave the non-poisonous but not-very-tasty red berries entirely to the birds.

Another advantage is that this deciduous bush needs little attention and can tolerate shade. This means that it is fine situated in hedges, especially in the neighbourhood of other shrubs that are good for feeding birds. In the wild, the species appears in forests.

TIP: The few fruiting varieties of Alpine currant, such as Majorenhof and Frankfurt, are difficult to obtain. The fruits are sweet and are a valuable supplement to redcurrants.

Varieties
- Aureum: golden-yellow foliage, lower-growing than the species
- Frankfurt: fruiting variety, broad bush with plentiful sweet fruits
- Majorenhof: very good fruiting variety, 1–2 m (3.25–6.5 ft) high
- Pumilum: small bush, up to 1 m (3.25 ft) high
- Schmidt: more compact than the wild form, 0.80–1.8 m (2.6–5.9 ft) spread; few fruits but a good pollinator for the fruiting varieties

PLANT PROFILE
- Height 1.50–2 m (4.9–6.5 ft), spread 1–2 m (3.25–6.5 ft)
- Greenish flowers that stand upright, from April to May
- Red berries
- Thornless twigs
- Likes sun, partial shade and shade
- For a rich, not-too-dry soil that may be chalky
- Easy to propagate by seed

Wool carder bee *Anthidium manicatum*

Easy to lure into domestic gardens, this yellow-and-black-striped bee is a species with remarkable behaviour. While the average drones of other bee species lead a relatively peaceful life, the male wool carder bees guard their harem by the feed-plants, hovering in front of the flowers like small helicopters and pouncing on the females to mate with them. They also fearlessly attack bumblebees and other insects competing to visit the flowers, using the spines at the rear end of their abdomen as weapons.

The drones are considerably larger than the females. The latter line their nests, which they make in cavities above or below ground, with hair that they shave off plant stems. This is a solitary species; each female wool carder bee takes care of her nest on her own.

Despite their attacks on honey bees and bumblebees, the drones always behave perfectly peacefully towards humans. They are not affected by spectators, so you can happily observe them in action.

BEE PROFILE

- Male 14–18 mm (0.55–0.7 in), female 10–12 mm (0.39–0.47 in) long
- Abdomen black with yellow stripes or spots, thorax pale and hairy at the sides
- Pollen transported on the ventral hairs
- Flying period June to September in one generation
- Nests in available cavities, also in old furry bee nests

Favourite plants

- Betony (*Stachys officinalis*)
- Black horehound (*Ballota nigra*)
- Downy woundwort (*Stachys germanica*)
- Foxglove (*Digitalis purpurea*)
- Lamb's ear (*Stachys byzantina*)
- Lavender (*Lavandula*)
- Motherwort (*Leonurus cardiaca*)
- Pickerelweed (*Pontederia cordata*, for ponds)
- Purple toadflax (*Linaria purpurea*)
- Rose campion (*Silene coronaria*, for the hair)
- Sage (*Salvia officinalis*)
- Spiny restharrow (*Ononis spinosa*)
- Strawflower (*Helichrysum*, for the hair)
- Wall germander (*Teucrium chamaedrys*)
- Woolly foxglove (*Digitalis lanata*)
- Yellow woundwort (*Stachys recta*)

Support

Motherwort in particular quickly tempts wool carder bees into the garden. For nest building, the garden should offer a selection of perennials with hairy leaves and stems. If you haven't got much space, go for lamb's ears as it offers both nourishment and hairs. Quinces are also visited. Building a dry stone wall or a herb spiral will provide the bees with the cavities they need for nesting. The males will be also be happy to spend the night in bee hotels.

→ Scene from a marriage: wool carder bees mating.

→ (inset) Deadly weapons: spines at the rear end of the abdomen of a drone.

Motherwort *Leonurus cardiaca*

If there is a true flower for the wool carder bee, it is probably the native motherwort. No sooner had I planted mine in the garden – acquired especially for these yellow-and-black customers – than the bees arrived. To the uninitiated, this imposing perennial may look rather like a stinging nettle, but when the pink flowers appear it is undoubtedly superior. It certainly looks magnificent with the light behind it. Its robust nature and long flowering period mean it scores highly not only with wool carder bees but also with butterflies and bumblebees. It also combines well with other plants for wool carder bees. Its size recommends it for the back of a bed or the centre of an island bed.

TIP: Grasses or imposing sage species such as garden sage (*Salvia officinalis*) and the biennial clary sage (*S. sclarea*) fit well with the wild nature of motherwort. In former times, motherwort was often found together with the equally bee-friendly black horehound (*Ballota nigra*) beside paths and fences, but this cottage garden plant combination has since become rare.

PLANT PROFILE

- Height 50–150 cm (1.6–4.9 ft), spread 50 cm (1.6 ft)
- Pink flowers from June/July to September
- Likes sun or partial shade
- For rich, not-too-dry soils
- Undemanding medicinal plant
- Easily propagated from seed

Downy woundwort *Stachys germanica*

This sometimes short-lived perennial looks very similar to lamb's ears, but its coat of grey fur is not quite as thick. Wool carder bees like it, despite its thinner hair, and the pink flowers are also on its plunder list. So, if you value plants that don't migrate in your garden, this is the better option. It even copes with stony soil, but does not get as out of hand as its Asiatic relatives.

It has become quite rare in Central Europe and certainly deserves to be rescued and allowed to feature in the garden. As plants are not often available in garden centres, buy seeds and sow them in early spring.

TIP: In its natural position, downy woundwort likes to be surrounded by other proven drought-lovers. Recommended planting for reproducing this – for example when creating a low-maintenance lawn imitating dry grassland – is a combination of meadow sage (*Salvia pratensis*), clustered bellflower (*Campanula glomerata*), wild marjoram (*Origanum vulgare*), yellow alfalfa (*Medicago falcata*) and salad burnet (*Sanguisorba minor*). This will delight not only bees but also butterflies.

PLANT PROFILE

- Height 30–100 cm (1–3.25 ft), spread 30 cm (1 ft)
- Pink flowers from June to August
- For sun or partial shade
- Likes loose, chalky soil that doesn't get waterlogged
- Very drought tolerant

Betony *Stachys officinalis*

This ancient medicinal plant with its pink blossoms is a real magnet for insects. Bumblebees and bees arrive in hordes and they attract some butterflies, too. Depending on the garden environment and the region in which you live, you can reckon on seeing some brimstone butterflies. Just make sure that you plant plenty of these slender native plants so that the wool carder bees do not claim everything for themselves. What's more, this woodland perennial with its low flower stems and neat leaves looks attractive at the front of a bed or along a wooden fence, making humans happy, too.

TIP: This perennial is easy to grow from seed sown in the spring. The plants grow quickly and usually start flowering in the second year. Among my seedlings, I also found a plant with pale pink flowers. The seed heads are decorative and can be left standing until the following spring.

Varieties
- Alba: white flowers
- Hummelo: darker flowers that are almost violet
- Pinky: pink flowers

PLANT PROFILE

- Height 20–70 cm (8–28 in), spread 30 cm (12 in)
- Deep pink flowers from June to August
- For sun or partial shade
- For relatively poor clay soils
- Tolerates drought well
- Avoided by slugs
- Medicinal plant

Lamb's ear *Stachys byzantina*

This perennial, originating in the Near East, is like a well-stocked supermarket for the wool carder bee, providing wool from the silvery-haired leaves and food from the pink flowers that stand out in strong contrast to the grey inflorescence. The soft, cuddly plant has a thick furry coat, which makes it very undemanding as it can manage without much water. This means that lamb's ear is good ground cover for areas of full sun, where it spreads by means of short runners and rapidly creates a hunting ground for wool carder bees. It can be combined with other drought-tolerant plants that are suitable for wool carder bees, such as lavender and sages. Bumblebees and other wild bees love it, too. They also enjoy sucking the morning rain or dew from the hairy leaves.

TIP: This is a perennial that is very good for giving away if friends and neighbours ask you for offshoots from your garden when they have seen the plants in bloom. The plants available in garden centres are usually the popular variety Silver Carpet, which very seldom flowers. The varieties Big Ears and Cotton Boll also produce very few flowers.

PLANT PROFILE

- Height including inflorescence 10–40 cm (4–16 in), ground-cover plant
- Pink blossoms July to August
- Likes sun
- For poor dry soils that don't get waterlogged
- Hairy silvery leaves and stems
- Basal leaves last through the winter

Hairy-footed flower bee
Anthophora plumipes

This tubby little bee may be mistaken for a bumblebee but, unlike the latter, flower bees have a really hectic lifestyle. They are always in a rush, catching our attention with their high-pitched buzzing.

In spring, before the first worker bumblebees are on the move, the hairy-footed flower bee is very easy to recognise by its proboscis, which is extremely long compared to that of a queen bumblebee. The females are unusual in having either a brownish or a black coat. The brown drones are very active, forever circling the garden with brief intermediate stops at the food-plants, where they mate with as many females as possible. If they come across a rival at this time, they will sometimes intercept him in mid-air. The wild males then fall to the ground and continue fighting there until one of them gives up. You can even pick these fighting bees up in your hand; they are far too preoccupied to take any notice.

The females make their nests in the soil, in earth banks or dry stone walls, preferably in colonies. The males like to sleep in hollow stems or holes drilled in bee hotels, but the nesting aid must have compartments filled with fine soil so that the females will move in too and excavate nests.

⬇ Deteriorating elderly female on common comfrey.

↑ Siesta: a male sunning itself on a hyacinth. The hairy fringes on the middle leg can be clearly seen.

Favourite plants
- Aubretia (*Aubrieta deltoidea*)
- Bugle (*Ajuga reptans*)
- Common comfrey (*Symphytum officinale*)
- Creeping comfrey (*S. grandiflorum*)
- Dead-nettles (*Lamium* spp.)
- Early-flowering borage (*Trachystemon orientalis*)
- Fumewort (*Corydalis solida*)
- Ground ivy (*Glechoma hederacea*)
- Holewort (*Corydalis cava*)
- Late tulip (*Tulipa tarda*)
- Lungwort (*Pulmonaria*)
- Oxlip (*Primula elatior*)
- Rough comfrey (*Symphytum asperum*)
- Wallfower (*Erysimum cheiri*)

Support

You can support these lively bees with the right food-plants, especially in the spring, and by providing suitable arrangements in an insect nesting wall. For instance, you can fill wooden boxes with loose soil and pile them up with the earth showing at the front. The filling needs to be soft, so that the bees can excavate it. Boring a few short tunnels into it will make it easier for the bees to move in.

BEE PROFILE

- 14–15 mm (0.55–0.59 in) long
- Pollen transported on the hind legs
- Males reddish-brown with striking yellow faces and long fringes of hair on the middle legs, females brown or black; both are very hairy
- Flying period March to June in one generation; the males start flying a few weeks before the females
- Nests in tunnels in the ground that they excavate themselves, often in steep artificial slopes in a bee hotel
- The nest site may be more than 100 m (325 ft) from the feed-plants
- The nests may be parasitised by the mourning bee (*Melecta albifrons*) (page 94)

Creeping comfrey *Symphytum grandiflorum*

Although hairy-footed flower bees happily take note of any kind of comfrey, this perennial from the Caucasus is their first resort as it flowers in their main busy period, in April and May. The fact that comfrey produces runners and soon takes over whole areas certainly suits both flower bees and bumblebees very well, but it can be annoying for the gardener. For this reason, you should plant it in positions close to trees and bushes, where green plants do not flourish. Strongly growing neighbours can also help to keep the dominant, continuously flowering comfrey in check. Apart from this, you don't have to bother about it at all – it needs very little care.

TIP: The variety Blaue Glocken ('Blue Bells') is more restrained and less vigorous than the species. A good partner for this, and popular with bumblebees, is Solomon's seal (*Polygonatum multiflorum*), which will be at the top with its arched flower-stems and will grow through the comfrey.

Varieties
- Blaue Glocken: blue flowers, less vigorous than the species
- Hidcote Blue: blue and white flowers
- Miraculum: the flowers change colour from red to blue and then to white; at 40–50 cm (1.3–1.6 ft) it is taller than the species
- Wisley Blue: pale blue flowers

PLANT PROFILE
- Height 20–30 cm (8–12 in), ground-cover plant
- White flowers from April to May
- For sun and partial shade
- Can even tolerate dry spots under deciduous bushes and hedges, and can cope with root pressure
- For normal soils that aren't waterlogged
- Produces runners

Fumewort and holewort *Corydalis solida, C. cava*

To provide the flower bees and bumblebee queens with plenty of these small perennials takes time and humus-rich soil. However, these native plants grow larger over the years and seed plentifully everywhere, thanks to the ants that disperse the seeds over a wide area (page 16). So, fumewort and holewort will turn up all around the garden and eventually form thick carpets of pink or white flowers in early spring under deciduous bushes. After the seeds have ripened, the plants die down into their underground tubers, leaving the terrain open to others. Fumewort is the more robust of the two and spreads more freely – a trait that make it popular with plant breeders, as can be seen in some varieties.

TIP: Plant the tubers under hornbeam hedges, where they will provide flowery ground cover. Digging over the soil is very disruptive for these plants, so do nothing – leave the fallen leaves lying where they are – and you will be guaranteed a carpet of flowers under hedges.

Varieties
- *Corydalis solida* G P Baker: very dark red flowers
- *C. solida* Kissproof: sensational variety for collectors, white flowers with blue lips
- *C. solida* Purple Bird: larger than the species, plentiful flowers
- *C. solida* Beth Evans: plentiful pretty pink blossoms

PLANT PROFILE
- Height and spread 15–30 cm (6–12 in)
- Pink, rarely white, scented flowers from March to April
- For places under deciduous bushes and hedges that are shady in summer but catch the sun in spring
- For humus-rich, chalky soils, similar to woodland
- Does not tolerate deep annual digging
- Self-seeding

Lungwort *Pulmonaria*

Hairy-footed flower bees love all kinds of lungwort, whose calyxes they can easily reach with their unusually long proboscises. Bumblebee queens and the first butterflies of the year also like to visit these perennials. Chameleon-like, the flowers gradually change colour from pink to blue, making the plant look many-coloured. An added bonus is the white-spotted leaves, which are so strongly marked in some varieties that they look quite silvery, meaning that these varieties can serve as decorative foliage plants throughout the year. One of the available species of lungwort (*Pulmonaria officinalis*) is widely naturalised in the UK.

TIP: Another favourite of flower bees that goes well with lungwort is the oxlip, whose pale yellow flowers provide an attractive contrast. However, early-flowering borage, which is taller and a bit of a bully, should be kept at a distance from the delicate lungwort.

Species and varieties:

- *Pulmonaria officinalis* Dora Bielefeld: pale pink flowers that don't change colour, forms carpets
- *P. officinalis* Ice Ballet: white flowers, long flowering period
- *P. officinalis* Wuppertal: vigorous variety, more abundant flowers than the species
- *P. rubra* Redstart: reddish flowers in large numbers, mildew resistant and vigorous
- *P. saccharata* Opal: pale blue flowers, long flowering period; summer leaves silvery
- *P. saccharata* Reginald Kaye: leaves almost white
- *P. officinalis* Alba: white flowers

PLANT PROFILE

- Height and spread 25–40 cm (10–16 in)
- Flowers first pink, then blue, from March to May
- Prefers partial shade or shady positions under deciduous shrubs
- Likes not-too-dry, chalky soil with plenty of humus; grows particularly well in loamy soils
- Grateful for added compost and a little leaf litter
- Seeds itself and spreads via runners

Early-flowering borage — *Trachystemon orientalis*

This large-leaved perennial from Eastern Europe is my secret tip if you like borage but find growing it annually from seed too much trouble or don't have enough space in the sun. Early-flowering borage has very similar flowers that appear before the leaves, and it is hardy in winter. At any rate, you should allow it plenty of space as its huge 25 × 15 cm (10 x 6 in) leaves that resemble elephants' ears, and a root system that never stops expanding, make it the top dog for shady problem zones in the garden. Here, it even forms a reliable bastion against ground elder. I use it to conceal the dark patch between compost bins and climbing roses, where it patiently ekes out its shadowy existence. Under deciduous shrubs with enough sunlight at flowering time, both flower bees and bumblebees find this a great food source.

TIP: These plants are easy to grow from seed, but the seedlings are sensitive to two degrees or more of black frost, so they may need to be protected in winter in the first year. The leaves and stalks of this ruffian are prepared as a vegetable in Black Sea cuisine.

PLANT PROFILE

- Height 20–40 cm (8–16 in), ground-cover plant
- Blue flowers from March to April
- For partial shade and shade
- Likes humus-rich soil that can be dry or even slightly moist
- Grows happily under shrubs
- Moderately self-seeding
- Avoided by slugs

Davies' colletes — *Colletes daviesanus*

This small wild bee with its golden fur and narrow-ringed abdomen loves plants of a sunny nature, such as tansy and yarrow. The creatures are so tiny that a female just has room to sit on a tansy bud. It shares its favourite plants with the bare-saddled colletes (*Colletes similis*), which is a less frequent visitor to gardens. Davies' colletes is grateful for nesting aids and moves in very quickly if its food-plants are in a nearby bed.

Favourite plants
- *Achillea filipendulina*
- Chamomile (*Matricaria chamomilla*)
- Common yarrow (*Achillea millefolium*)
- Feverfew (*Tanacetum parthenium*)
- Tansy (*T. vulgare*)
- Yellow chamomile (*Anthemis tinctoria*)

Support
You can encourage the bees with artificial slopes of light soil. Wooden boxes filled with this soft material can easily be integrated in larger nesting walls with the filling at the front. If you get a good take-up rate, you can add extra modules. If you do not have space for wooden boxes, try hanging space-saving clay nest blocks in the sun as an inducement to lure the bees in.

BEE PROFILE
- 7–9 mm (0.27–0.35 in) long
- Reddish-brown hairy thorax and head, abdomen black with a whitish band at the end of each segment
- Pollen transported on the hind legs
- Flying period June to August in one generation
- Nests that they excavate themselves in mortar, sand or loam – preferably in steep slopes; they will also settle in ready-made holes in clay nesting aids
- The nests are sealed with a transparent membrane

Achillea filipendulina

This Caucasian yarrow is an easy-care alternative to tansy. Both species are bright yellow and have an aromatic scent. The bees are happy to lay claim to the immigrant as a pollen source. This perennial with its large flower heads and imposing presence suits both cottage gardens and modern gravel gardens, where it is in good company with red valerian, sedum, sage and wallflowers. The good news is that achillea is not a thirsty plant. It just needs a position with enough sun; if it is in too much shade, fruit flies will be forever successfully sabotaging the flowers.

TIP: Cutting back promptly encourages a second flowering. The flower stems can be left standing over the winter. Snow-capped or garnished with hoar frost, they are a welcome sight when you're drinking your morning coffee.

Varieties
- Altgold: compact, up to 60 cm (2ft) high
- Cloth of Gold: one of the tallest varieties, up to 1.6 m (5.25 ft) high, strong with large flowers
- Coronation Gold: large flowers
- Golden Plate: up to 1 m (3.25 ft) high, stable variety
- Parker: strong and robust

PLANT PROFILE
- Height 60–120 cm (2–4 ft), spread about 50 cm (1.6 ft)
- Flat, golden yellow flower heads from June to September
- For sunny spots, but can also tolerate partial shade
- For normal to fairly dry garden soil, which should be nutrient rich
- Aromatic plant that slugs won't touch

Yellow chamomile *Anthemis tinctoria*

This perennial, which ensures a happy mood in flowerbeds and meadows, is distinguished by its yellow, marguerite-like flowers. It looks very pretty and is equally uncomplicated: it puts up with the driest and stoniest places and never complains, as long as it is in the sun. This delicate native plant will also not run wild and is rarely nibbled by snails. The only thing it can be reproached for is its short life expectancy. However, if the plant has enough free space around it, it seeds itself.

Yellow chamomile is visited by beneficial hoverflies and butterflies as well as by the smaller wild bees. In addition to Davies' colletes, *Heriades* and bare-saddled colletes (*Colletes similis*) may appear on the scene. The leaves are eaten by the larvae of the chamomile shark moth (*Cucullia chamomillae*).

TIP: The plant will live longer if you cut it back after flowering. Then it will flower again and delight the bees all the longer. Always allow a few flowers to go to seed so the plant can propagate itself.

Varieties
- Ala Dagh: Anatolian dyer's chamomile, relatively long-lived
- Lemon Ice: cream-coloured selection with yellow centres
- Sauce Hollandaise: white with yellow centres – like small marguerites

PLANT PROFILE

- Height about 30–60 cm (12–24 in), spread 30 cm (12 in)
- Yellow flowers from June to September
- Prefers a position in full sun, also partial shade if need be
- Likes poor, dry, even stony soils
- Aromatic plant

Tansy *Tanacetum vulgare*

The pretty tansy is a native meadow plant that is passionately adored by the very smallest of the wild bees. In addition to several types of plasterer bee, its sunny yellow flowers are visited by *Heriades* and the yellow-faced bee *Hylaeus nigritus*. The larvae of many butterfly species eat its aromatic ferny foliage. However, despite its undemanding nature and long flowering period, tansy is not too popular among gardeners because it is inclined to spread uninhibitedly. It causes fewest problems in a flower meadow or a spacious island bed in the lawn.

TIP: The old variety Crispum with its very frilly leaves is rather more modest in its growth. It is so exclusive that it is more likely to thin out than to spread. Nevertheless, if you don't have a very natural garden or a restricted space where tansy can be allowed to romp about, it would be better to switch to this tame yarrow. The aforementioned bees will be just as pleased to visit it and even the decorative larvae of the chamomile shark will eat both species.

PLANT PROFILE

- Height 1–1.2 m (3.25–4 ft), spread 60 cm (1 ft)
- Yellow flower heads, like small buds, from July to September
- For sunny spots
- Likes rich soils, which shouldn't be too dry
- Produces both runners and seeds
- Aromatic plant
- Good flower for cutting

Ivy bee *Colletes hederae*

Unlike the honey bee, the ivy bee catches the eye with the fine rings on its abdomen and the thick brown hairs on its thorax. Its late flying period from September on and its predilection for the inconspicuous flowers of ivy make this wild bee easy to spot, because most wild bees avoid ivy. The ivy bee, by contrast, has specialised in ivy as a source of pollen and would on no account miss its flowering period.

The males start flying before the females and have to bridge the gap with other plants. They do this by visiting all possible flowers, such as calamint (*Clinopodium nepeta*; see photo) and mountain fleece (*Polygonum amplexicaule*). Although ivy is widespread, the bee appears only in warm places. Although a relatively new arrival in the UK, it is now found regularly in the south of England and in Wales.

Favourite plant
- Common ivy (*Hedera helix*)

Support
Ivy that is old enough to flower should be kept, the bigger the better. New plantings are an investment in the future of this bee and may in time increase existing populations. If the bees have settled in your sandpit, the children should not play in it in September and October. Providing a substitute for the bees in the form of an area of sandy soil is a better solution.

BEE PROFILE

- Male 10 mm (0.39 in) long, female up to 14 mm (0.55 in) long
- Abdomen yellow and brown striped, thorax brownish and hairy
- Unlike the honey bee, it folds its wings together on its back when visiting flowers
- Pollen transported on the hind legs
- Flying period September to October in one generation
- Excavates nests in the ground, often in sandpits or lawns with thin grass
- The brood cells are coated with a silk-like substance

Common ivy *Hedera helix*

This evergreen native is the rescue service when it comes to prettifying ugly walls or fences and making them non-transparent. It climbs up walls and tree trunks with the aid of adventitious roots. After about eight years, ivy forms adult, non-climbing growths that are noticeable because the shape of the leaves is different. Flowers are produced on this shrubby growth only. You can even weave juvenile stems into a wire mesh fence. When planning this, take account of the future weight and the volume of the adult growth, which may stand out a good metre. It is worth planting, even without ivy bees: butterflies, hoverflies and honey bees will visit the flowers and birds like to eat the black berries.

TIP: If you have spotted ivy bees nearby but you haven't got a free corner for a vertical support, simply treat the insects to a hedge of shrub ivy. This will not climb and will flower soon after planting, and will mean that you will be able to observe the bees in comfort.

Varieties
- Arborescens: shrub ivy, up to 1.5 m (4.9 ft) high
- Arbori Gloss: slender shrub ivy with shiny foliage
- Aureovariegata: yellow variegated foliage, climbing
- Ice Cream: dwarf shrub with white variegated foliage
- Ovata: oval leaves, does not climb as high as the species, up to 8 m (26.2 ft)
- Zorgvlied: abundant flowering shrub ivy, only 1 m (3.25 ft) high

PLANT PROFILE
- The wild form can climb up to more than 20 m (65 ft) if allowed to
- For sun, partial shade or shade
- Late, yellowish flowers in September and October
- Undemanding, as long as there is humus in the soil
- The poisonous black berries ripen in winter
- Evergreen
- May grow very old

Sweat bees

Halictus species, *Lasioglossum* species

Sweat bees are mostly small, but they are great opportunists. The more common species that also appear in private gardens enjoy many different flowers, no matter how tiny they are; they can be found on both Virginia creeper and heuchera. Unlike many other bees, they do not scorn beetroot flowers.

Some sweat bees fly throughout the summer as, like bumblebees, they live in small colonies in nests that they excavate in the soil themselves. The individual types are hard to distinguish and are often overlooked because of their small size. What they all have in common is the lengthwise fork on the hindmost segment of their abdomen.

Favourite plants
- Asteraceae, such as dandelion or hawksbeard (*Crepis*)
- Cranesbill (*Geranium*)
- Cultivated and wild roses with single flowers (*Rosa*)
- Heuchera
- Perennial cornflower (*Centaurea montana*)
- Wild marjoram (*Origanum vulgare*)
- Willow (*Salix*)

Support
Because of their long flying period, sweat bees benefit from a garden where there is always something in flower. They are content with plants that are of peripheral importance to the flowerbeds and can be encouraged particularly by the shoots of vegetables such as Swiss chard and beetroot. Avoid deep digging as this will disturb the nests. Areas with poor vegetation or small slopes of loose soil where they can dig will provide additional help for the bees.

→ Sweat bee at work on purple tansy.

↓ Intimate detail: close-up of the fork in the rear of the abdomen.

BEE PROFILE
- Only 4–16 mm (0.16–0.63 in) long depending on species; the male *Lasioglossum* is also noticeably slender, with long antennae
- Pollen transported on the hind legs
- Flying period mostly from April into autumn
- Nests in the ground, which it excavates itself, often communally or even in small, annual colonies

Yellow-faced bee *Hylaeus* species

They look like Zorro – shiny black, with a pale mask that fills almost the entire face of the males and is made up of smaller markings in the females. All the species of yellow-faced bees are small and very similar to each other in appearance, making them difficult to identify. However, some give themselves away by their specialised choice of flowers. For example, the onion yellow-face (*Hylaeus punctulatissimus*) spurns all garden plants except for summer-flowering alliums. The mignonette yellow-face (*H. signatus*), on the other hand, has a weakness for plants of the *Reseda* family. Yet another species, *H. nigritus*, likes tansy and its cultivated form, *Achillea filipendulina*. The common yellow-face (*Hylaeus communis*) is less choosy and is consequently the one most frequently found in gardens.

↑ Geranium Rozanne with a yellow-faced bee.

→ Yellow-faced bees like to visit mignonette, a member of the *Reseda* family.

Favourite plants
- Chives (*Allium schoenoprasum*)
- Cranesbill (*Geranium*)
- Fennel (*Foeniculum vulgare*)
- Giant allium (*Allium giganteum*)
- Leek (*A. porrum*)
- Mignonette (*Reseda odorata*)
- Round-headed leek (*Allium sphaerocephalon*)
- Small yellow onion (*A. flavum*)
- Weld (*Reseda luteola*)
- White mignonette (*R. alba*)
- Wild mignonette (*R. lutea*)

Support
A bee hotel with hollow stems is always popular with yellow-faced bees, and they would be very pleased to have the chance to visit flowers in a nice old-fashioned cottage garden that also has vegetables and herbs growing in it. If there are leeks flowering there, so much the better; otherwise ornamental alliums are a good addition. The forgotten members of the mignonette family that are usually grown as annuals or biennials would not only please the bees but will also be a desirable way to halt their decline.

BEE PROFILE
- Only 6–9 mm (0.23–0.35 in) long, depending on the species; almost uniformly shiny black with a pale face
- No external pollen transport; instead, the pollen is swallowed together with the nectar
- Flying period from May to September in one or two generations
- Often makes its nest in old beetle tunnels in wood, so also happy in a bee hotel; other species nest in brambles
- The nest is sealed with a characteristic transparent membrane

Leaf-cutter bees
Megachile ericetorum, M. nigriventris

These are the rodents among wild bees. They are called leaf-cutters because they line their nest tunnels with oval pieces of leaf, which they bite out of rose leaves, for example, and assemble in small packets – in each of which lives a larva with its food supply of pollen. Despite their surgical operations on the foliage of garden plants, they are not pests. Other *Megachile* species that are difficult to distinguish come into gardens. You can best observe these bees in summer on leguminous plants, such as sweet peas.

Favourite plants
- Bladder-senna (*Colutea arborescens*)
- Broad-leaved everlasting pea (*Lathyrus latifolius*)
- False indigo (*Baptisia australis*)
- Lamb's ear (*Stachys byzantina*)
- Lavender (*Lavandula*)
- Spiny restharrow (*Ononis spinosa*)
- Sweet pea (*Lathyrus odoratus*)

Support
Legumes are a must, as are mints such as betony. These leaf-cutter bees are usually pleased to see all kinds of nest site, from dry stone walls to bee hotels and dead wood.

BEE PROFILE
- *Megachile ericetorum* (see photo): 10–14 mm (0.39–0.55 in); *M. nigriventris* (page 46): 12–16 mm (0.47–0.63 in) long
- *M. nigriventris* looks like a teddy bear because of the light brown fur on its thorax
- *M. ericetorum* is easily recognisable by the striking transverse stripes on its abdomen
- Pollen transported on the ventral hairs, grains often visible on top of the abdomen when collecting
- Flying period June to August
- *M. nigriventris* nests in dead wood
- *M. ericetorum* makes its nests in available cavities, including bee hotels and dead wood

Bladder-senna *Colutea arborescens*

This tirelessly flowering legume is a reliable partner for leaf-cutter bees. Butterflies also visit this shrub, and some blue butterflies even use the leaves as nurseries. The flowers are in warm summery colours of yellow or orange and are a spectacle in themselves, but the plant is by no means content with that. The bladder-senna continues flowering all summer, while forming countless inflated fruits whose semi-transparent cases shine like lanterns when they are lit from behind. Inside, there are black seeds ripening, waiting for the wind to carry off the entire bladder and find a new home somewhere for the precious fruit. However, most of the bladders remain on the bush until the winter. In summer, the fruits and flowers are around at the same time, giving this easy-care shrub an exotic and colourful look.

TIP: Where I live in the city, the bladder-senna is often cut hard back in February. It looks shocking every time, but the shrubs quickly put out new branches that will flower again in June. So if the bush becomes too big, you can simply cut it back in this way.

PLANT PROFILE

- Height 2.5–3 m (8.2–9.8 ft), spread 1–2 m (3.25–6.5 ft)
- Yellow or orange flowers from May until about September
- Dense, deciduous shrub without thorns
- For sunny or part-shaded positions
- Tolerates dry, poor soils that may be chalky
- Both seeds and leaves are poisonous

False indigo *Baptisia australis*

This robust North American perennial with its splendid blue flowers is quite rarely planted. It is one of the acknowledged favourites of leaf-cutter bees. In the photo it is pictured with *Megachile nigriventris*. Bumblebees also romp around it. The spring shoots look almost like green asparagus; then the plant becomes bushy with bluish-green foliage. It is very drought tolerant and is rarely troubled by pests. By collaborating with nitrogen-fixing bacteria, false indigo can cope with poor soils. The pods containing the black seeds turn bluish as they ripen, and when flowering is over they still look attractive in winter. The seed germinates easily, but a few years will elapse before the seedlings can match up to the mother plant.

TIP: Unfortunately this perennial does not flower a second time if it is dead-headed, so spare yourself the effort and instead enjoy the pods, which beneficial earwigs like to use after ripening to conceal the roofs of their nests.

Varieties:
- *B. australis* var. minor: 0.7 m (2.3 ft) high
- Blue Pearls: 1.3 m (1.25 ft) high, with blue and white flowers

PLANT PROFILE
- Height 0.6–1.5 m (2–4.9 ft), spread 0.6 m (2 ft)
- Blue flowers from June to July
- Long-lived perennial with attractive foliage
- For sunny to part-shaded positions
- Copes with poor, dry positions, and also sandy soil
- Not invasive
- Dye plant

Broad-leaved everlasting pea *Lathyrus latifolius*

It looks like hard labour and weightlifting when leaf-cutter bees squeeze their way into the flowers of the everlasting pea. The pistil is pushed upwards and brushes the sides of the bees. The effort involved is too great for honey bees, so they bite off the flowers. However, the violet carpenter bee (*Xylocopa violacea*) is strong enough to deal with everlasting peas. Bumblebees, brimstone butterflies and many moths, including the hummingbird hawkmoth, also visit the flowers.

This climbing perennial makes a beautiful screen for the summer, flowering tirelessly from June on and needing very little watering. The tendrils need a stable, 2 m (6.5 ft) high frame to climb up, otherwise they have nothing to hold on to. If their way upwards is barred, they grow horizontally and climb over other plants.

TIP: Cut off the fruits to encourage the perennial to continue flowering. Although the plant is robust, its stems break easily; they can't be forced to grow the way you want them to and need to be tied in at an early stage.

Varieties
- Rosa Perle: pink flowers with very thin darker stripes
- Rote Perle: pink flowers that turn bluish as they go over
- Weisse Perle: brilliant white flowers

PLANT PROFILE

- Height 1–2.5 m (3.25–8.2 ft), spread about 1 m (3.25 ft) over the years
- Pink flowers from June to August
- Perennial plant for sunny to part-shaded positions
- Undemanding with regard to soil, drought tolerant
- Almost ignored by snails
- Very long-lived
- Easy to grow from seed

Red mason bee and European orchard bee

Osmia bicornis, O. cornuta

These two species, especially the red mason bee (*Osmia bicornis*), will give you a real sense of achievement. They will quickly discover a new nesting aid with tunnels in wood or hollow plant stems. However, they are also extremely inventive when it comes to finding unconventional niches, such as garden hoses, dowels or holes in wooden shelving. They even happily accept holes that run vertically.

The European orchard bee (*O. cornuta*), which looks like a small red-tailed bumblebee with its rusty-red abdomen and black thorax, also definitely likes warmth and prefers big-city gardens. Mason bees are eager to help with the pollination of fruit trees, but otherwise they are not fussy.

The males of both species start flying before the females and wait impatiently outside the nesting sites for a swarm of females. You can easily pick out paired insects near the nesting aid by their loud, rhythmic buzzing. Mated females that have long been busy nest building will continue to be approached and annoyed by bachelors. They lay out the nest chambers one behind the other, fill them with pollen, lay an egg and seal each chamber with a wall of moist earth. In larger cavities, the nests are also laid out in a higgledy-piggledy fashion.

⬇ European orchard bees engaged in family planning.

BEE PROFILE

- *Osmia bicornis*: 8–15 mm (0.3–0.59 in); *O. cornuta*: 12–16 mm (0.47–0.63 in) long
- The abdomen of *O. bicornis* is rusty-red for the most part, with black hairs only on the last segments; the thorax is light brown; the males have paler hair on their faces and are more delicate
- *O. cornuta* has a black hairy thorax and a reddish abdomen; males have white faces
- Both species have small horns on the clypeus below the antennae
- Pollen transported on the ventral hairs
- Flying period March to June
- Nests in available cavities, often in bee hotels

↑ Red mason bees love scillas.

Favourite plants
- Apple (*Malus*)
- Aubretia (*Aubrieta deltoidea*)
- Cherry plum (*Prunus cerasifera*)
- Crocus (*Crocus*)
- Grape hyacinth (*Muscari*)
- Hawthorn (*Crataegus monogyna*)
- Pear (*Pyrus communis*)
- Siberian squill (*Scilla siberica*)
- Wallflower (*Erysimum*)
- Willow (*Salix*)
- Winter heath (*Erica carnea*)
- Winter-flowering cherry (*Prunus subhirtella*)

Support
A bee hotel with bamboo stems and wood with ready-bored holes will be happily adopted. Clay nesting aids are also successful. The diameter of the holes should be 5–9 mm (0.2–0.35 in). As the nests are made with moist earth, a garden pond where the insects can collect building material at the muddy edge is a great help. The soil in the cracks of a rough stone wall is also sought after.

Solitary mason bee *Osmia rapunculi*

Although this bee is so small, it gives me the greatest pleasure when I walk around my garden and see one sleeping in a bellflower. The reddish-haired drones also catch attention with their beautiful green eyes. The females gather food for the larvae exclusively from bellflowers, whereas the males also like cranesbill.

Luckily, there are bellflowers to suit many situations in life – from climbing species for walls and containers to ones that can cope with drought and full sun, and others that prefer to grow in partial shade. If you have a bee hotel as well, your garden will become the focus of life for this specialised bee.

Favourite plants
- Clustered bellflower (*Campanula glomerata*)
- Giant bellflower (*C. latifolia*)
- Harebell (*C. rotundifolia*)
- Nettle-leaved bellflower (*C. trachelium*)
- Peach-leaved bellflower (*C. persicifolia*)
- Rampion bellflower (*C. rapunculus*)
- Trailing bellflower (*C. poscharskyana*)

Support
These bees like to settle in a nesting aid that contains hollow stems such as bamboo or reeds with an internal diameter of 3–4 mm (0.12–0.16 in). You can provide the insects with various species of bellflower. The males are happy with cranesbill flowers, but for a peaceful night's rest they will want a bellflower to sleep in.

BEE PROFILE
- Females 8–9 mm (0.31–0.35 in), males 9–11 mm (0.35–0.43 in) long
- Very slender, males with noticeably curved abdomen
- The females are covered in thin grey hairs, the males have reddish-brown hairs
- Pollen is transported on the ventral hairs
- Flying period June to August
- Nests in available cavities, such as in horizontal hollow stems in a bee hotel

Clustered bellflower Campanula glomerata

This native perennial is far too seldom planted, yet its clusters of bells make it perfect for cutting – if you have the heart to deprive the bees of a few flowers. When selecting blooms, be sure to watch out for any bees that might be sleeping in the flowers. Fortunately, there is never a lack of new flowers as the clustered bellflower produces runners. This plant is a playground for bumblebees and honey bees as well as the solitary mason bee.

Varieties
- Acaulis: small, only 15 cm (6 in) high
- Alba: white flowers, 40 cm (16 in) high
- Caroline: pink flowers, often starting as early as May, up to 50 cm (20 in) high; more slow-growing than the species
- Dahurica: imposing variety, up to 60 cm (24 in) high
- Joan Elliott: very large flowers, 40 cm (16 in) high
- Superba: up to 60 cm (24 in) high, abundant flowers

TIP: Clustered bellflowers can spread quite widely in a flowerbed if they have no competitive neighbours to keep them in check. However, if planted in meadows and natural areas or on sunny slopes, the large varieties in particular – such as Dahurica and Superba – can be very effective. The wild species will be included in good wildflower seed mixtures for creating flower meadows.

PLANT PROFILE
- Height 30–60 cm (12–24 in), spread 40 cm (16 in)
- Bluish-purple or white flowers from June to August
- Likes sun or partial shade
- Wild species grow in meadows with poor soil; in the garden it tolerates humus-rich soils that may be chalky
- Tolerates drought
- Produces runners

Trailing bellflower *Campanula poscharskyana*

This bellflower with the unpronounceable botanical name is perfect for turning your garden upside down. It climbs down from walls or troughs, but it can grow equally well under roses or in a rock garden, where its masses of blue flowers will be very satisfying. Even in a balcony box you can use it to attract solitary mason bees and other species. The pollinating activities of the bees will ensure you get even more plants as the trailing bellflower is happy to distribute seedlings, even in cracks in walls if need be. If it is growing in a container in strong midday sun, this Southern European perennial will sometimes need to be watered, but otherwise it is easy to care for and even remains green in winter.

TIP: I planted this bellflower in a large container more than 10 years ago, as ground cover for an Alpine clematis (*Clematis alpina*); since then it has provided blue flowers from May to September. I add a little compost to the pot in the spring, but no other fertiliser.

Varieties
- Blauranke: vigorous, with long tendrils
- Blue Crown: blue flowers with white centres, slow-growing
- Stella: slow-growing, with flowers of a slightly darker blue
- Templiner Teppich: low ground-cover plant, largely avoided by snails

PLANT PROFILE
- Height 15–20 cm (6–8 in), ground-cover plant
- Blue flowers from June to July and again in September
- Prefers sun or light shade
- Does not like to be waterlogged, can cope with poor soil
- Evergreen and seeds itself easily

Giant bellflower *Campanula latifolia*

When the flowers of my neighbour's white climbing rose hang over into my garden and the giant bellflowers are blooming abundantly below them, I am just as delighted as the many visiting insects. Bumblebees, yellow-faced bees, leaf-cutter bees, mining bees and, of course, the solitary mason bee (*Osmia rapunculi*) love the large flowers of this tall perennial. The blue-flowered varieties occasionally produce specimens with white flowers. However, the native plant not only fits in well with roses; it can also make do with a less sunny area at the edge of shrubs. The seed heads can be left on the plant, guaranteeing plenty of seedlings – after all, you never know when the voles will strike and nibble the tubers of the giant bellflower.

TIP: In my garden, the solitary mason bees are rather conservative, preferring the blue flowers to the white. With aphids it seems to be the other way round, so I give the blue-flowered seedlings preferential treatment in order to have more bees.

PLANT PROFILE

- Height around 100 cm (3.25 ft), spread 50 cm (1.6 ft)
- Blue or white flowers from June to July
- Prefers sun, partial shade or light shade under deciduous shrubs
- For rich soils with sufficient moisture
- In summer drought the plant dies down after flowering and sends up new shoots the following year
- Seeds itself; does not produce runners
- Protection against voles is advisable, for example a wire-mesh basket

Large-headed resin bee
Osmia truncorum

Resin bees are with us all summer long, tirelessly building new nests in the bee hotel and sealing the entrances with resin. They are so industrious that I sometimes wonder whether they might in fact annex other bees' cells, throw out their neighbours' pollen and then bring in provender for their own larvae!

While it would be impossible to provide enough nesting opportunities to meet the insatiable needs of these little black insects, all that is actually required for a large colony of resin bees to grow is a good supply of nearby food-plants in the form of *Asteraceae*.

Favourite plants
- *Achillea filipendulina*
- Common yarrow (*A. millefolium*)
- Elecampane (*Inula helenium*)
- Tall fleabane (*Erigeron annuus*)
- Tansy (*Tanacetum vulgare*)
- Telekia (*Buphthalmum speciosum*)
- Yellow chamomile (*Anthemis tinctoria*)

Support
If the bee hotel is fully booked by mason bees flying in May, it is helpful to hang up another new nesting aid in June with holes 3–3.5 mm (0.12–0.14 in) in diameter. The resin bees will quickly move in and have a colony of their own. They love to relax a bit in the garden and are delighted by many wild plants, such as groundsel, chicory, chamomile and cat's ear. The pretty native tansy is particularly recommended.

BEE PROFILE
- 4–8 mm (0.16–0.32 in) long
- Small bee, almost black with a few grey hairs, a stocky body and a very large head
- Pollen transported on the ventral hairs
- Flying period mid-June to September
- Nests in beetle tunnels in dead wood, also happy in a bee hotel in drilled holes and hollow plant stems
- Nests are sealed with resin with tiny stones worked into it

Elecampane *Inula helenium*

Of the many attractive, yellow-flowered plants for resin bees, the mighty elecampane has made it on to the short list as its always rather unkempt-looking flowers are also visited by butterflies and bumblebees. All the same, this large-leaved perennial, as tall as a man, is not exactly modest and needs a lot of space in all directions, so it is best kept at the back of the bed. Its long flowering period in high summer means it can also be recommended as a single plant, perhaps in a special island bed of its own. It is easy to grow from seed, but you can also propagate it by dividing the huge rhizome.

TIP: If the elecampane should get too exuberant, you can use the root for all kinds of experiments – it can be used for dyeing, as a cough medicine and even as an air freshener. If, despite all that, you have too little space for this large perennial, it would be better to switch to the delicate but shorter-lived yellow chamomile (*Anthemis tinctoria*), which also tolerates dry positions.

PLANT PROFILE

- Height 1.8–2 m (5.9–6.5 ft), spread 1 m (3.25 ft)
- Yellow daisy-like flowers with a diameter of around 7.5 cm (3 in) from July to August, often into September
- European perennial for sun and partial shade
- Likes rich soils, not too dry, with a high proportion of humus
- Medicinal plant; seeds itself easily

Violet carpenter bee
Xylocopa violacea

Carpenter bees are real fatties that catch the attention with their loud buzzing and hummingbird-like flight. Despite their self-assured appearance and their imposing size, these black bees with blue-tinged wings are very peaceful. However, you will only come to truly enjoy these giants, which fly from spring into late summer, if you live in a region with a warm climate. If not, the species may nevertheless soon visit your garden as climate change means that it is sure to become more widespread. Prepare a warm reception for it by offering a wide range of flowers throughout the season and plenty of rotten dead wood, in which the females can gnaw out their tunnels.

Favourite plants
- Aubretia (*Aubrieta deltoidea*)
- Bear's breeches (*Acanthus*)
- Broad-leaved everlasting pea (*Lathyrus latifolius*)
- Butterfly bush (*Buddleia davidii*)
- Clary sage (*Salvia sclarea*)
- King's spear (*Asphodelus luteus*)
- Lamb's ear (*Stachys byzantina*)
- Wisteria (*Wisteria*)

Support
Carpenter bees need plenty of dead wood in sunny situations. Whole tree trunks or parts of trunks, for instance of fruit trees, are particularly suitable. These bees even nest in the fruiting bodies of tinder fungi, which they collect and arrange in a dry place with dead wood.

BEE PROFILE
- 2–2.5 cm (0.79–1 in) long
- Black-haired, shiny bodies, metallic blue wings
- Pollen transported on the legs and in the crop
- Flying period March to September in one generation
- Chews its nests in soft dead wood

Long-leaved bear's breeches *Acanthus hungaricus*

The imposing bear's breeches (*Acanthus mollis*) from the Mediterranean area is more often seen carved in stone than in gardens, as its leaves adorn many ancient columns. However, as it is slightly frost sensitive, it is better to choose long-leaved bear's breeches (*A. hungaricus*) for the garden. It is the hardiest species and, thanks to its habit of growing in clumps, it is also the easiest to control.

This perennial is a very stylish adornment to the flowerbed when the large flower stems grow up from the decorative rosette of prickly leaves. The lower lip of each flower is whitish, while the upper part is pink with purple veins. Carpenter bees and bumble bees are just the right size for the roomy flowers. Stems that have finished flowering remain decorative for a long time and add structure to the bed.

TIP: The warmth-loving bear's breeches is suitable for rough situations, preferably with a wall at its back to protect it from damp in winter. You can even keep both types of bear's breeches in containers, if you protect them from winter frosts.

PLANT PROFILE

- Height 60–120 cm (2–4 ft), spread 70 cm (2.3 ft)
- For sun or partial shade
- Pink and white flowers from July to August
- For well-drained soils in beds or along hedges
- Does not produce runners

Chinese wisteria *sinensis*

Everything about this Asiatic climber is gigantic: the long racemes of blue flowers form scented curtains on pergolas and walls and their twining tendrils with large leaves can cover several storeys of a house at once. Wisteria is a twining climber and so strong that it can even crush downpipes and other pliable parts of buildings like straws, so it needs a very stable support. The side shoots can be cut hard back in high summer. Carpenter bees and bumblebees are very fond of the large flowers.

TIP: Wisteria can be underplanted with other blue flowers, such as bluebells. If you can't provide the wisteria with a suitable support to climb up, you don't have to go without it as it is possible to buy plants that have been grown as trees. You can also train young specimens yourself with the aid of a sturdy supporting stake.

Varieties
- Alba: white flowers
- Caroline: starts flowering after only a few years
- Prolific: vigorous with very long racemes, climbs up to 10 m (32 ft)

PLANT PROFILE

- If not cut back, the plant will reach 6–10 m (19.7–32.8 ft) in height and up to 8 m (26.2 ft) in spread
- For sun or partial shade
- Blue, scented flowers in May and June, often a second time in August
- For a rich light soil, with little chalk and not too dry
- Doesn't tolerate waterlogging
- Warmth loving
- Poisonous seedpods
- Deciduous, turns yellow in autumn

Clary sage *Salvia sclarea*

The biennial clary sage invests in a rosette of leaves in the first year, then in the following summer it surpasses itself, putting many perennials in the shade. Above its large leaves appear pink and white flowers surrounded by colourful bracts to give greater effect at a distance. These bracts provide colour from the moment the flower buds appear. This sage looks particularly splendid with the light behind it, and in addition the plant gives off a spicy aroma, which means that the leaves can be used as a herb. Clary sage is visited mainly by large, powerful bees, such as bumblebees, wool carder bees and carpenter bees.

TIP: Good neighbours requiring similar positions and equally little watering are lavender, common sage and Russian sage. Leave the seed heads standing. If the plant has sufficient space around it and it is in a sunny position, it will generally seed itself.

Variety
- *Salvia sclarea* var. *turkestanica* Vatican White: white flowers and bracts

PLANT PROFILE

- Height 0.8–1.2 m (2.6–4 ft), spread 0.6–1.2 m (2–4 ft)
- For sun or partial shade
- Pink flowers from June to August
- Biennial
- Likes dry, stony or sandy soils
- Self-seeding
- Aromatic plant

Bumblebees
Furry colonisers

Bumblebees are everybody's darlings. Always warmly wrapped up, they also help out in the garden in bad weather, when other bees prefer to stay at home. This chapter presents the four most commonly found urban species and the best plants for them.

Common carder bee
Bombus pascuorum

The common carder bee is my representative for dealing with vegetables and is always keen to take care of the tomato flowers. It is one of the most common bumblebee species, with a long flying period lasting into autumn. The secret of its success is its extremely flexible choice of nesting sites. The nests may be above ground in tufts of grass or cushions of moss as well as in mouse nests, hollow trees, old birds' nests in nest boxes, and even in buildings. Common carder bees are extremely peaceful creatures, so they are perfectly suited to settling in nest boxes, and often move in.

Favourite plants
- Early-flowering borage (*Trachystemon orientalis*)
- Flowering currant (*Ribes sanguineum*)
- Ice plant (*Sedum*)
- Lungwort (*Pulmonaria*)
- Perennial cornflower (*Centaurea montana*)
- Spotted dead-nettle (*Lamium maculatum*)
- White dead-nettle (*L. album*)
- Wild teasel (*Dipsacus fullonum*)

⬇ The common carder bee, a bumblebees with a long proboscis, seen on green alkanet (*Pentaglottis sempervirens*).

Support
The common carder bee likes to see many flowers at all times of year. Because it often nests above ground, a little bit of wilderness with big tufts of grass is perfect for it, and you can also offer it plenty of smaller, attractive, clump-forming, ornamental grasses.

> **BEE PROFILE**
> - The queen is 15–18 mm (0.59–0.7 in) long, the workers are 9–15 mm (0.35–0.59 in) and the drones are 12–14 mm (0.47–0.55 in)
> - Colour variable, with light brown thorax, blackish-grey abdomen with reddish-brown band at the end, older insects sometimes greyer
> - Long flying period from April to October
> - Fairly small colonies of 60–150 bees
> - Long proboscis

Tree bumblebee *Bombus hypnorum*

The increasingly common tree bumblebee is a real little drama queen and the only species of bumblebee that occasionally stings when it's having a bad day. The workers can be irritable, especially in reaction to sultry weather. The stormy atmosphere is quickly transferred to these insects, mainly because they build their nests above ground in bird boxes, hollow trees and on buildings, which often get too hot and, on top of everything else, are too small for a large colony.

A hot-tempered mood is easily recognised by the loud buzzing when the insects are ventilating their nests. I have had tree bumblebees as my neighbours on my balcony for several years in succession, and they have remained very peaceful because of the shady location. However, if it were to come to stinging matches around the nest box, many beekeepers will be prepared to move the whole colony and its lodging to a less contentious site.

Favourite plants
- Apple (*Malus*)
- Blackberry (*Rubus fruticosus*)
- Jostaberry (*Ribes* × *nidigrolaria*)
- Wild rose (*Rosa*)

Support
Tree bumblebees often move into bee boxes, which they easily find for themselves. As they prefer holes in trees with old birds' nests in them, you can also help by having largish pieces of dead wood in the garden.

⬇ Tree bumblebee on dusky cranesbill.

BEE PROFILE
- The queen is 16–20 mm (0.63–0.79 in) long, the workers 8–18 mm (0.3–0.7 in) and the drones 14–16 mm (0.55–0.63 in)
- Thorax red-brown on top, abdomen black with white band at the tail
- Flying period very short, from March to August
- Colony size up to 400 insects
- Bumblebee with short proboscis that prefers to visit open flowers

Buff-tailed bumblebee

Bombus terrestris∎

This is probably the largest, best-known and most beautiful species of bumblebee, with the neat yellow stripes in its black fur, which looks less tousled than that of many of its relatives. In spring, the large queens catch the eye as they meticulously search the garden floor for mouse holes, inspecting every dark opening. This also includes compost heaps, where mouse nests can often be tracked down. If the rodent is still in residence, it will be driven out by force of arms by using the sting.

These bees visit a large number of plant species, but because of their short proboscises they prefer open flowers and break into flowers with long corolla tubes.

Favourite plants
- Aquilegia (*Aquilegia*)
- Blue eryngo (*Eryngium planum*)
- Common poppy (*Papaver rhoeas*)
- Crocus (*Crocus*)
- Holewort (*Corydalis cava*)
- Iris (*Iris*)
- Lavender (*Lavandula*)
- Purple coneflower (*Echinacea purpurea*)
- Purple tansy (*Phacelia tanacetifolia*)

Support
Buff-tailed bumblebees prefer to nest in mouse holes, either underground or in compost heaps. However, as the latter are usually dismantled in spring when the flowerbeds are bare and the compost will have a warming effect on the soil, the bees cannot easily move into a compost bin unless you empty it in autumn instead. The queens will settle in nest boxes, but rarely discover them for themselves.

⬇ Buff-tailed bumblebee on stinking hellebore.

BEE PROFILE

- The queen is 20–23 mm (0.79–0.9 in) long, the workers 11–17 mm (0.43–0.67 in) and the drones 14–16 mm (0.55–0.63 in)
- Black with two yellow bands and a buff tail to the abdomen
- Flying period from mid-March to August
- Colony size 100–600 insects
- Short proboscis

Garden bumblebee *Bombus hortorum*

This bee has the longest proboscis of all the bumblebees – around 15 mm (0.59 in) in length for both workers and drones. Having a proboscis of this size is a bit of a hindrance with hollyhocks, but for flowers with a longer calyx tube it is extremely useful. That is why you also find the garden bumblebee on typical butterfly plants, such as honeysuckle, which are avoided by the similar-looking buff-tailed bumblebee with its shorter proboscis. Another noticeable difference from the buff-tailed bumblebee is the three yellow stripes, two of which border the join between the thorax and the abdomen. The drones look more yellow.

Favourite plants
- Comfrey (*Symphytum officinale*)
- Common foxglove (*Digitalis purpurea*)
- Honeysuckle (*Lonicera periclymenum*)
- Iris (*Iris*)

BEE PROFILE
- The queen is 18–26 mm (0.7–1 in) long, the workers 11–16 mm (0.43–0.63 in) and the drones 13–15 mm (0.51–0.59 in)
- Black with three yellow bands and a white tip to the abdomen
- Relatively long head
- Flying period from mid-March to August; young queens sometimes produce a second generation in the year of their hatching, and these may continue flying into October
- Colony size 50–150 insects
- Long proboscis

- Monk's hood (*Aconitum napellus*)
- Small yellow foxglove (*Digitalis lutea*)

Support
Garden bumblebees are not very choosy and are as happy to nest in mouse holes as in birds' nests, barns and attics. They are also completely open-minded about bee boxes. Young queens often find their own way back to the same nest box the following year.

⬇ Garden bumblebees have a long proboscis.

Aquilegia   *Types of aquilegia*

Aquilegias are short-lived perennials that remain in the garden because they are freely self-seeding. They are also a help with garden design; wherever they pop up, I always feel that it is exactly where they are needed. They even sow themselves in gaps between paving stones, where their taproots become rather two-dimensional.

Bumblebees adore their blue, purple or pink flowers. The aquilegia is actually designed to make the insects approach the flowers from below in order to collect pollen and pass it on to other plants. However, as the spurs on the blossoms are very long, only bumblebees with long proboscises can reach the nectar. The ones that draw the short straw simply chew a hole in the spurs. Buff-tailed bumblebees are masters at breaking into flowers and thus also provide direct access to the nectar for honey bees and small sweat bees.

TIP: Aquilegias cannot be divided because of their taproot. However, that does not matter as the seeds germinate readily and the young plants can be used at an early stage to fill gaps in flowerbeds. One very handy thing about these perennials is that snails largely ignore them. Only voles may become a problem.

PLANT PROFILE

- Height 50–60 cm (1.6–2 ft), spread 30 cm (12 in)
- Likes sun or partial shade
- Blue, pink, white or lilac flowers from May to June
- For rich, not-too-dry soils, *Aquilegia vulgaris* is very drought tolerant
- Don't cut off the seed heads before the seeds are ripe, so they can self-seed
- Avoided by snails

Korean mint *Agastache rugosa*

Botanists have shown great imagination and given this perennial several appropriate names for safety's sake, such as Korean mint, Indian mint, purple giant hyssop and blue liquorice. Although the plant has travelled far, our native bumblebees, hoverflies and butterflies flock to its stiff, upright blue flower stems, which provide food for months on end. A combination of white- and blue-flowering varieties creates colourful summer beds adorned with bright butterflies. Because of its long flowering period, this Asian perennial is particularly recommended for small gardens.

TIP: The seed heads are also attractive throughout winter, especially in hoar frost. They remain firmly upright and add structure to the garden, so it is worth waiting until the spring before cutting them back.

Varieties
- Alabaster: white flowers
- Alba: also white
- Black Adder: large hybrid with dark flowers and a long flowering period
- Blue Fortune: hybrid, strong and compact form
- Golden Jubilee: yellowish foliage, blue flowers

PLANT PROFILE

- Height 60–120 cm (2–4 ft), depending on the variety, spread 40 cm (16 in)
- Blue flowers from July to September, until October if the weather is mild
- Perennial plant for full sun
- Tolerates dry, sandy soil, but also any normal garden soil that doesn't get waterlogged
- Forms clumps but not invasive
- Aromatic and medicinal plant

Russian sage *Perovskia atriplicifolia*

This is an *éminence grise* of imposing stature, with bright blue flowers in high summer. A subshrub from the Asian steppes, it takes problematic dry positions in its stride, as long as it doesn't get its feet too wet in winter. If planted in spring in well-drained soil with the addition of sand or gravel, Russian sage is reliably hardy in winter and flowers its heart out even in the hottest summers; it is scented as well. Bumblebees aren't the only creatures to take advantage of the flowers – sparrows like to peck out the ripened seeds, even while it is still in flower.

TIP: In terms of position requirements, Russian sage goes well with lavender, valerian, catmint, lamb's ear and common sage. If that is a bit too much grey foliage for you, combine it with the purple coneflower (*Echinacea purpurea*) and achillea (*Achillea filipendulina*).

Varieties

- Blue Spire: with a height of 100–150 cm (3.25–4.9 ft) it is one of the most imposing selections
- Filigran: 80–120 cm (2.6–4 ft) high, compact with many flower stems
- Lacey Blue: one of the smallest varieties, up to 50 cm (20 in) high
- Little Spire: a compact variety, 70–80 cm (2.3–2.6 ft) high, also suitable for large containers

PLANT PROFILE

- Height 70–150 cm (2.3–4.9 ft), depending on variety, spread 100 cm (3.25 ft)
- Aromatic subshrub with blue flowers from August to September
- Sun-loving
- For dry, chalky or sandy soils that don't get waterlogged

Turkish sage *Phlomis russeliana*

Rosettes of unobtrusive, pastel-yellow flowers around the stem are the speciality of the Turkish sage. The stems add height to the flowerbed, even when the flowers are over. This robust perennial has a long flowering period, continually adding yet another rosette and starting a new tier at the top. The flowers have hoods that only powerful insects can raise in order to reach the pollen and nectar. Bumblebees, especially buff-tailed ones, are happy to make the effort and squeeze their way in.

Over time, Turkish sage can build up a large number of plants by means of runners, so only position it next to other perennials that can hold their own.

TIP: Don't cut down the seed heads until late winter as they look decorative throughout the season and marsh tits search them for seeds. The large seeds germinate easily and can be used for quick propagation, as well as cutting off runners.

Alternative
- Jerusalem sage (*Phlomis tuberosa*): rosettes of pink flowers, up to 1.2 m (4 ft) high; for dry, sunny soils

PLANT PROFILE
- Height 30–100 cm (12 in–3.25 ft), ground-cover plant
- Yellow flowers from June to July
- Evergreen perennial
- Tolerates sun, partial shade and light shade around the edges of shrubs
- Rich soils that don't get waterlogged
- Spreads via short runners
- Slugs will not touch it

Monk's hood *Aconitum napellus*

The place where you can most rely on being able to admire this splendid perennial in the wild is in the Alps, where the bumblebee *Bombus gerstaeckeri* has specialised in feeding on this and related species. However, more northern bumblebees love the blue flowers, too. Gardeners who grow monk's hood and are plagued by snails can heave a sigh of relief. Its poisonous flowers and leaves are largely avoided by these pests, giving it a great advantage over delphiniums.

It is undemanding with regard to choice of position and grows well even in partial shade around shrubs or in light shade. My plant is growing at the foot of an ornamental apple tree. The seed heads are also attractive and last a long time.

TIP: Fancy a bonus? The blue flowers of the Asian Carmichael's monk's hood (*Aconitum carmichaelii*) follow on seamlessly from those of its relative, flowering from September. It too is poisonous. If the frost doesn't arrive until late, flowering will continue into November and delight the late bumblebees, especially the queens. Honey bees also like to visit it.

Varieties
- Bicolor: blue and white flowers
- Gletschereis: brilliant white variety
- Kleiner Ritter: the name means 'little knight' and it is appropriately compact and sturdy, up to 90 cm (2.9 ft) high
- Rubellum: pink flowers
- Schneewittchen: white variety, up to 1.5 m (4.9 ft) high

PLANT PROFILE
- Height 50–150 cm (1.6–4.9 ft), spread 40 cm (16 in)
- Blue flowers between June and August
- Native perennial for sun, partial shade or light shade under deciduous shrubs
- Does not like too-dry or rich soils
- Large specimens often need support
- Poisonous plant
- No danger from snails

Common foxglove *Digitalis purpurea*

This is not a plant for impatient people. In the first year the foxglove forms a splendid rosette of leaves, and in its second summer it sends up a spectacular flower stem with blossoms that open from the bottom upwards and last for a long time.

Garden bumblebees in particular love this flower, although it doesn't make things easy for them. They have to fly up from below into the long tubular flowers, whose interiors are as slippery as a children's slide. Many foxgloves play a bit of a joke by developing a gigantic flower like a satellite dish at the very top. If the bumblebees have been working hard, the plant will make masses of fine seeds, which you can scatter over empty spaces in the flowerbed. These germinate easily.

TIP: Foxgloves like to disappear after flowering, but there are things you can do to extend their life. Immediately after the last flower has withered, cut the seed head off at the base. Then the plant will forget that it has already flowered and will live for a further year. Seeds will already have formed below, so seeding is assured.

Other species

- Large yellow foxglove (*Digitalis grandiflora*): perennial with yellow flowers from June to July, for sun to light shade
- Small yellow foxglove (*D. lutea*): native perennial, sometimes biennial, with yellow flowers from June to August; for dry, part-shaded positions
- Woolly foxglove (*D. lanata*): biennial with white and gold flowers from July to August; likes dry, sunny spots; also suitable for wool carder bees

PLANT PROFILE

- Height 80–100 cm (2.6–3.25 ft), spread 30 cm (12 in)
- Pink or white flowers from June to July
- Native biennial for sun to part shade, often around woodland edges
- For a rich soil that shouldn't be too chalky
- Self-seeding
- Poisonous plant

Perennial cornflower *Centaurea montana*

Cornflower-blue flowers are the trademark of this grey-leaved perennial, which originated in the mountains of Central Europe but has long been established in lowland cottage gardens. The fat buds produce a sweet secretion that is particularly popular with ants from glands known as 'extrafloral nectaries'. Then, in early summer the plant becomes the darling of the bumblebees. Common carder bees are the most frequent visitors to these flowers, which readily give up their nectar and pollen. After pollination the fat seeds ripen in a silvery casing. Goldfinches are often on the spot before this, preventing the plant from seeding itself by nibbling the nourishing fruits.

TIP: If you fancy more extravagance in your flowerbeds, there is good news: unusual varieties are now available, with flowers in various shades, from white to almost black, and with even longer flowering periods. But be warned: you are in danger of becoming addicted. Say hello to collecting fever.

Varieties
- Alba: white flowers with pink centres from May to July
- Amethyst in Snow: white flowers with violet centres from May to July
- Black Sprite: almost black flowers from May to July
- Carnea: pale pink flowers from May to July
- Gold Bullion: yellowish foliage
- Grandiflora: larger flowers than the species, will flower again in September if cut back
- Merel: purple flowers from May to July

PLANT PROFILE
- Height 30–50 cm (12–20 in), spread 40 cm (16 in)
- Blue flowers from May to June
- Likes a sunny or part-shaded position, also woodland edges
- For humus-rich soil, tolerates drought
- Self-seeding, and spreads widely

Honeysuckle *Lonicera periclymenum*

This native climbing honeysuckle with long corolla tubes is actually a typical plant for butterflies and moths. Its main customers are moths, which is the reason why the flowers switch on their beguiling scent in the evening. Privet hawk moths, gamma moths and humming-bird hawk moths are regulars, the last two also in the daytime. The garden bumblebee with its extra-long proboscis also gets to enjoy the flowers. As the plant blossoms tirelessly, it deserves a place in the bumblebee garden, preferably near a seat, where you can enjoy the scent. The berries are popular with songbirds.

TIP: My honeysuckle on the rose arch often has a bad hair day in summer as the upper tendrils twine only around themselves. After a few experiments, I have now got the hang of things and I cut the honeysuckle back at the end of February and in mid-June, so that it is always closely attached to the top of the supporting arch. Although this delays flowering a little, the blossoms are all the more abundant on compact growth.

PLANT PROFILE

- Height 2–6 m (6.5–19.7 ft), spread up to 2.5 m (8.2 ft)
- Sunny to part-shaded position
- Flowering period from June to the first frost
- Prefers not-too-dry, rich loamy soils
- Shiny red, poisonous berries in clusters
- Deciduous, leaves reappear early
- Needs a stable support at least 2 m (6.5 ft) high, with plenty of vertical struts for the tendrils to twine around
- May be under-planted with perennials
- Spring shoots can be used for propagation by cuttings

Obedient plant *Physostegia virginiana*

This North American perennial, also known as obedience or false dragon head, is not immediately impressive. Its tall flower stems always stay nicely upright, while the obedient plant slowly improves the soil by means of its shallow, creeping root system. In my garden it rescued a sunny, waterlogged place in a bed. Where other perennials committed mass suicide, the obedient plant was the first to flourish.

You can safely turn the head of this plant in any direction; each individual flower sits on a joint, so you can bend it to the left or the right and it will stay there. Bumblebees appreciate the late, foxglove-like blooms. Humming-bird hawk moths are also guaranteed to look in.

TIP: Good planting partners for a late summer bumblebee bed are Culver's root (*Veronicastrum virginicum*) and the Japanese anemone (*Anemone hupehensis*). The shallow blooms of the anemones are popular with bumblebees and are plentiful. Only snails take no pleasure in this combination.

Varieties

- Bouquet Rose: deep pink
- Crystal Peak White: white flowers from July to September, compact variety
- Summer Snow: white flowers
- Vivid: late variety, flowering from September to October

PLANT PROFILE

- Height 70–90 cm (2.3–3 ft), spread 30 cm (1 ft)
- Pink or white flowers from August to September
- Prefers a sunny or part-shaded position
- Likes rich, slightly damp soils
- Spurned by snails

Bergamot *Monarda didyma*

Although the well-known lemon balm is much in demand among bumblebees, it can become a nuisance because of its many offspring. The North American bergamot is much easier to control. You can use its aromatic leaves for teas and as a flavouring. The wild species is particularly good for this. It has dark red, very long-lasting flowers that are visited by bumblebees and butterflies. Many wild bees also make use of the flower heads as a place to sleep. In some varieties, the colour of the flowers is supplemented by reddish bracts.

This perennial combines particularly well with grasses and giant hyssops (*Agastache*). Equally recommended for the bumblebee garden is the closely related and very similar-looking wild bergamot (*Monarda fistulosa*), which has many pink-flowered varieties. Spotted bee balm (*M. punctata*) with its pink bracts is suitable for sandy soils.

TIP: Clumps of bergamot gradually become wider over the years, but at some stage acquire a bald patch in the interior. Then it is time to get out your spade and divide the plant, which will keep it vigorous and flowering readily.

Varieties
- Alba: white flowers
- Fireball: bright red, small variety, only 40 cm (16 in) high
- Jacob Cline: very large red flowers
- Marshall's Delight: tall variety with pretty pink flowers
- Petite Delight: pink flowers, only 30 cm (12 in) high
- Squaw: large red flowers, which are resistant to mildew

PLANT PROFILE
- Height 70–150 cm (2.3–4.9 ft), spread 50 cm (1.6 ft)
- Red flowers from June to August
- Prefers sun or partial shade
- For a rich, not-too-dry soil
- Not threatened by slugs
- Aromatic and culinary plant

Bastard balm *Melittis melissophyllum*

Because it stays nicely in place and does not scatter runners everywhere around itself like the dead-nettles, the bastard balm has to be propagated from cuttings or seed. That can take a long time, especially as each flower produces only four seeds, so this native perennial is seldom available to buy. That makes it just right for an exclusive bumblebee garden.

The plant likes to be left in peace in light shade around shrubs and can grow very old. It is also suitable for planting in a shady rock garden or in the shadow of a dry stone wall, as it likes the warmth. The crushed leaves smell of honey. The flowers are quite large for such a small plant, and are white with a red underlip.

TIP: Bumblebee-friendly partners for the bastard balm are the stinking hellebore (*Helleborus foetidus*) and the spring pea (*Lathyrus vernus*) – both native perennials that like chalky, stony positions in partial shade under shrubs.

Varieties
- Album: pure white flowers
- Royal Velvet Distinction: bicoloured flowers, vigorous, clump-forming

PLANT PROFILE

- Height and spread 30–50 cm (12–20 in)
- Pink and white flowers from May to June
- Suitable for sun, light shade and partial shade around shrubs
- For rich, chalky soils that contain some humus
- Likes stony subsoil that meets the plant's need for warmth
- Foliage turns dark in sunny positions
- Aromatic plant

Faassen's catmint *Nepeta × faassenii*

Many cats love this grey-leaved perennial, but we plant it more for other furry creatures – bumblebees. They are very fond of this blue labiate and value its long flowering period, which can be extended by cutting back after the first flowering. Honey bees and butterflies are always around it as well.

In addition, this plant is impressive because of its undemanding nature and tolerance of drought. Catmints never push their way into the foreground, but they are reliable and persuasive with their clouds of blue flowers. Tall forms can be used to hide the bare feet of roses, small varieties can make an unobtrusive edging to a bed. As long as it can look on the sunny side of life, a catmint cannot be intimidated – except perhaps by an infatuated cat.

TIP: Catmints are particularly attractive if you add a few balls to their aromatic haze of flowers. With their globular flower heads, alliums that flower in early summer – especially the ornamental onion (*Allium aflatunense*) – provide a beautiful picture for bees as well.

Varieties
- Dropmore: 40–60 cm (1.3–2 ft) high, very robust
- Gletschereis: only 50–60 cm (1.6–2 ft) high, white flowered
- Six Hills Giant: 60–80 cm (2–2.6 ft) high, goes well with roses
- Snowflake: 25 cm (10 in) high, low, white
- Walkers Low: 70–90 cm (2.3–2.9 ft) high, robust, vigorous

PLANT PROFILE
- Height 25–90 cm (10 in–3 ft), depending on variety, spread 40–60 cm (1.3–2 ft)
- Blue flowers from May to July, and again in September if cut back
- For a sunny or part-shaded position
- Likes quite poor soils that don't get waterlogged and may be dry
- Aromatic plants
- Largely avoided by slugs

Viper's bugloss *Echium vulgare*

If the native viper's bugloss were not so short-lived, it would certainly have a big fan club. Unfortunately, the plant is usually biennial, but luckily it maintains its presence through self-seeding. Being a true pioneer plant, it makes use of open spaces where there is no competition, such as may be found in gravel beds or other gravelled places. It gets over drought without a fuss.

Its pink flowers, which fade into blue, appear in large numbers over a long period of time on imposing flower stems and are popular with both bumblebees and butterflies. If you ever manage to get the viper's bugloss mason bee (*Osmia adunca/Hoplitis adunca*) to settle in your garden, you can be proud. This specialised bee needs a whole array of viper's bugloss to persuade it to stay.

TIP: Viper's bugloss is suitable for a dry bed with equally short-lived plants, as there will always be a free spot here for it. I recommend mullein (*Verbascum*), rose campion (*Silene coronaria*), hoary alyssum (*Berteroa incana*) and yellow chamomile (*Anthemis tinctoria*).

PLANT PROFILE

- Height 60–80 cm (2–2.6 ft), spread 40 cm (16 in)
- Flowers pink, then blue, occasionally white, from June to September
- Likes sun
- For poor, dry positions that don't get waterlogged; not for heavy clay soils
- Biennial or short-lived perennial, self-seeding
- May be grown in large containers

Balm-leaved archangel *Lamium orvala*

Dead-nettles, especially the spotted dead-nettle (*Lamium maculatum*) and white dead-nettle (*L. album*), should be in every bumblebee garden and they often move in by themselves. They like to live along hedgerows and around shrubs and meander around under the bushes. The king of them all is the balm-leaved archangel, also known as the giant dead-nettle, from the mountains of Eastern Europe. It is truly majestic with its tall, shrubby growth and large flowers, and it is a favourite of bumblebees and other insects. This plant is not invasive, so you can place it in a herbaceous bed, along with other woodland-edge plants. It is long-lived and tolerates a shady position close to the roots of shrubs.

TIP: Other plants that create an attractive contrast in form of growth and are also good for bumblebees are Solomon's seal (*Polygonatum multiflorum*) and snowy woodrush (*Luzula nivea*), whose grassy stems and white flowers add accents. In part-shaded positions the ornamental cow parsley (*Anthriscus sylvestris* Ravenswing) adds a touch of lightness to the planting and goes well with the dark flowers of the balm-leaved archangel.

Variety
- Album: white flowers

PLANT PROFILE

- Height 40–60 cm (1.3–2 ft), spread 40 cm (1.3 ft)
- Brownish-red flowers with pale pink lips from May to June
- For partial shade or shady positions
- Likes rich soils, tolerates drought

Common bugloss *Anchusa officinalis*

This native, usually biennial plant keeps the bumblebees on a string. More and more new flowers keep unfurling on upright flower stems. The relatively small flowers are first violet, later turning dark blue, but always with a white centre. Very occasionally, plants appear that have purple or completely white flowers. In the year of sowing, the plant develops a rosette of hairy leaves; the flowers appear the following summer.

Because of its taproot, the common bugloss grows happily on dry soils and even on sand. It can be combined with mignonettes, poppies and viper's bugloss to create a colourful flower meadow, much to the delight of bumblebees and butterflies; several butterflies lay their eggs on its leaves.

TIP: You can easily grow common bugloss (*Anchusa officinalis*) from seed sown between March and April. Once it is established, it will keep going by self-seeding. The bright blue garden anchusa (*A. azurea*) is slightly longer-lived but more sensitive to frost.

PLANT PROFILE

- Height 50–100 cm (1.6–3.25 ft), spread 50 cm (1.6 ft)
- Flowers first violet then blue, from June to September
- Likes a position in sun or partial shade
- For a poor, dry soil that doesn't get waterlogged
- Biennial, may sometime last for several years
- Self-seeding

Japanese rose *Rosa rugosa*

This species from Eastern Asia is the only wild rose that flowers throughout the summer. New enormous, scented flowers appear non-stop above shiny, healthy leaves, while at the same time the large hips are ripening and being enjoyed by the greenfinches. Bumblebees dive enthusiastically into the pollen-laden blooms that can be reliably found on the bushes. With luck you will also be able to spot bee beetles or rose chafers (see photo) if you don't trim off the old flowers.

TIP: A hedge of alternately planted pink and white bushes will make you dream of seaside holidays. You can very easily find a use for the vitamin-rich hips in the kitchen.

Varieties
- Alba: white flowers, otherwise like the species
- Dagmar Hastrup: pale pink flowers, 1 m (3.25 ft) high
- Foxi: violet-pink flowers, 50–80 cm (1.6–2.6 ft) high
- Moje Hammerberg: semi-double, pink flowers
- Schnee-Eule: white, semi-double flowers, 50–80 cm (1.6–2.6 ft) high
- White Roadrunner: semi-double white flowers; for small hedges, only 50 cm (1.6 ft) high and wide

PLANT PROFILE
- Height 1–2 m (3.25–6.5 ft), spread 0.5–1.5 m (1.6–4.9 ft), tolerates cutting
- Pink or white scented flowers from June into autumn, 6–8 cm (2.4–3.1 in) in diameter
- Orange to red hips
- Lovely autumn colour
- Many straight thorns
- Likes sun to partial shade
- For well-drained soils that shouldn't be chalky but rather acidic, also tolerates poor sandy soils
- Salt tolerant
- Forms tendrils, making it suitable for stabilising slopes

Common sage *Salvia officinalis*

All sages are a playground for bumblebees, but there is a good reason why this little subshrub with aromatic leaves is also known as culinary sage; it is frequently used in Mediterranean dishes. In addition, the range of varieties leaves no wish unfulfilled and has a stock of extravagant colours. Common sage likes a sunny position with well-drained soil. Its evergreen – or rather ever-grey – foliage can also withstand cold winters, as long as its roots are not too wet. As well as bumblebees, the blue flowers attract butterflies and wool carder bees. The caterpillars of the pretty gamma moth occasionally eat the leaves.

TIP: Together with hyssop (*Hyssopus officinalis*), thyme (any species of *Thymus*) and marjoram (*Origanum vulgare*), sage makes a bee-friendly kitchen garden that butterflies will also like.

Varieties
- Berggarten: very large, roundish leaves, but fewer flowers than the species
- Crispa: pretty, crinkled leaves
- Icterina: colourful yellow foliage
- Mittenwald: compact growth, very aromatic
- Nana Alba: dwarf form with white flowers, well suited to containers
- Purpurascens: purple foliage
- Rosea: pink flowers
- Tricolor: white, grey and purple variegated leaves

PLANT PROFILE

- Height 40–80 cm (1.6–2.6 ft), spread at maturity up to 1 m (3.25 ft)
- For a sunny position, but likes protection from winter sun
- Blue flowers from May to July
- For poor dry soils that mustn't get waterlogged
- Thrives best on stony subsoil
- Seeds itself on open ground
- Easy to grow from cuttings

Welsh poppy   *Meconopsis cambrica*

This yellow-flowered poppy is the shade-tolerant answer to the common poppy. Bumblebees and other bees find it equally desirable, but it is less labour-intensive as this species finds its own niches, without the need to break up the soil each year. The plants are perennial but not very long-lived. However, abundant self-seeding means that it will remain in the garden for as long as there is free space available. It provides highlights in the bee garden because of its long flowering period; new buds continue to open alongside the ripening seed capsules, which resemble little turrets.

TIP: Consider the Welsh poppy as a creative influence in the garden, forever conquering new favourite spots. If it gets to be too much of a good thing, weed it out quickly.

Varieties
- Aurantiaca: orange variant
- Frances Perry: almost red

PLANT PROFILE

- Height 30 cm (12 in), spread 20–30 cm (8–12 in)
- Prefers to grow in partial shade or light shade
- Yellow flowers from June to September, in mild winters often until the first frost
- Likes to live in loamy soils but also in drier areas
- For planting along hedges and walls or under light shrubs; will even germinate in chinks and crevices
- Likes it cool
- Short-lived perennial
- Not threatened by snails
- Native to Western Europe

Cup plant *Silphium perfoliatum*

This tall-growing North American perennial is the shooting star among energy plants and could replace the boring maize, which would be warmly welcomed by bees. However, you do not need a whole field in order to bring these plants into your garden – it also looks good at the back of a bed or screening a fence. The sunflower-like blooms appear over a large part of high summer, during which the food slowly goes out to the bumblebees. The leaves are unusual, growing in a similar way to those of the wild teasel to form a cup around the stem in which rainwater collects, and giving the plant its common English name.

TIP: The cup plant can be grown from seed, but not as easily as sunflowers. For the most successful germination, sow on open ground in autumn. By contrast, sowing in spring on a warm window-ledge produces poor results, as I have proved for myself. However, I had beginner's luck and one seed nevertheless graciously consented to germinate. As this perennial grows fast, a single plant is sufficient to begin with.

PLANT PROFILE

- Height 2–3 m (6.5–9.8 ft), spread 1 m (3.25 ft)
- Yellow flowers from July to September
- For sun or partial shade
- Prefers rich soils that can be moist but mustn't get waterlogged
- The plant remains considerably smaller on dry soils (under 2 m/6.5 ft)
- Self-seeding
- Doesn't produce runners

Purple gromwell   *Lithospermum purpurocaeruleum*

This native perennial is the altogether trouble-free package for light shade under shrubs or on dry slopes that you would like to leave to their own devices. Given the chance, the gromwell will soon take over the space intended for it and form a green carpet by sending out long, arching tendrils that root where they touch the soil. The leaves are pointed and covered with rough hairs. Its numerous flowers are at first pink but quickly turn blue. Both bumblebees and butterflies love these flowers. After pollination, it produces shiny white seeds as hard as stone that look almost as if they are made of porcelain. This means that the plant remains decorative after flowering, with a glittering stock of extravagant pearly beads.

TIP: Have you still got a space in a large container? The gromwell can be used here for underplanting small shrubs. Allow its shoots to hang attractively over the edge of the pot.

PLANT PROFILE

- Height 30–50 cm (12–20 in), ground-cover plant
- Flowers at first reddish then turning to blue, from April to June
- Likes sun, semi-shade or light shade, especially around shrubs or hedges
- Tolerates poor to slightly rich soils that may be dry and chalky
- Spreads quickly by layering

Hotspot plants
to set the garden buzzing

The plants in this chapter are real magnets for bees and there will always be something going on around them. They will either attract many individuals or many different species, sometimes even both – either way, it will be worth watching and listening.

Apple *Malus species*

While my miniature cherry is quite infuriatingly ignored by insects most years, an apple tree is always full of the joys of spring. No matter whether it is an ornamental or an eating apple, a profusely blossoming apple tree hums like a transformer. This sound is produced by honey bees, bumblebees and wild bees, especially mason bees and mining bees. So, a bee hotel housing plenty of mason bees can ensure a good harvest, even if there isn't a colony of honey bees nearby.

Unlike eating apples, the fruits of ornamental apples are no bigger than cherries and less tasty, but the birds enjoy eating them. Apple trees can easily be underplanted with perennials and flowering alliums. Old trees that no longer flower well should not be consigned to the scrap heap but instead be rejuvenated with the aid of a rambling rose. In addition, they often provide nesting holes for birds and rotten wood for wild bees to nest in.

TIP: Eating apples need a suitable pollinating variety nearby. If you are short of space, an ornamental apple can provide a remedy as the wild species can pollinate all other apples and are obtainable in almost all shapes and sizes.

PLANT PROFILE

- Height and spread depend on variety; anything is possible, from columnar trees to bush-shaped ornamental trees and tall eating apple trees
- Scented blossom in April or May: white, pink or lilac, depending on variety
- Likes a fertile, humus-rich soil, but basically undemanding

Marjoram *Origanum vulgare*

Nobody can pass this by. Marjoram is one of our most popular native perennials. Insects will abandon everything in order to dive into a cloud of pink, or sometimes white, flowers. There is plenty of time for them to do so, as its flowering period extends over several months. The purple perianths make the plant appear bicoloured.

Honey bees, wild bees, bumblebees, hoverflies and even butterflies are among its regular visitors. It also provides food for the caterpillars of a few species of butterflies and moths, most of which are so tiny that the traces of their nibbling seem quite discreet. In my garden a little jewel of an insect, the delicate mint moth (*Pyrausta aurata*), has regularly made this plant its nursery.

TIP: Despite its apparent delicacy, marjoram is a crafty fellow. It misses no opportunity to perpetuate itself throughout the garden by seeding. It is therefore advisable to cut it back after flowering. All the same, it makes up for its pushiness with its tasty leaves, which will liven up any pasta sauce.

Varieties
- Album: white flowers
- Compactum: only 20 cm (8 in) high, bushy and free-flowering
- Country Cream: white-variegated foliage
- Goldtaler: yellow leaves
- Thumbles Variety: green and yellow foliage, 25 cm (10 in) high; suitable for sunny positions

PLANT PROFILE
- Height 20–70 cm (8 in–2.3 ft), at times prostrate, spread 40 cm (16 in)
- Lilac-pink flowers between July and September
- Rosettes of evergreen leaves
- For sun, partial shade or light shade
- Also for poor dry soils, undemanding
- Vigorous perennial that seeds itself abundantly around the garden
- Culinary and medicinal plant

Culver's root *Veronicastrum virginicum*

This North American perennial sets off a truly explosive firework display of slender flower heads. The tall stems stand like exclamation marks in the bed, attracting hordes of insects. Honey bees and wild bees find their way to them, as well as hoverflies, bumblebees and butterflies.

The attractive foliage forms a star shape around the stem. The seed heads are also good to look at and should be left standing during the winter. Nevertheless, such splendour requires a lot of water; the plant will become stunted in dry soil.

TIP: The tall flower stems can be used in many combinations; for example, together with mountain fleece (*Polygonum amplexicaule*) they create an international firework display. You can create an all-American bed for bees with bergamots (*Monarda*), the orange coneflower (*Rudbeckia fulgida*) and the cup plant (*Silphium perfoliatum*).

Varieties
- Diana (see photo): white flowers, 1.2 m (4 ft) high
- Erika: dark buds, pale pink flowers, free-flowering, 1.5 m (4.9 ft) high
- Fascination: violet-blue flowers, often striped; up to 1.8 m (5.9 ft) high
- Lavendelturm: pale blue flowers, up to 1.9 m (6.2 ft) high
- Pink Glow: pale pink flowers, up to 1.4 m (4.6 ft) high

PLANT PROFILE
- Height 100–190 cm (3.25–6.2 ft), spread 30 cm (12 in)
- Stems of pale blue to whitish flowers from July to September
- North American perennial
- For sun and partial shade
- Likes rich, not-too-dry soils, which may also be fairly moist
- Not threatened by snails

Alder buckthorn *Frangula alnus*

The alder buckthorn flowers for months on end, which is much appreciated by hardworking bees, so there is always a buzz around it. After the small white flowers, it forms berries that change colour from red to black. Birds love these fruits, especially during the migration period in September. Alder buckthorn is also enticing for butterflies; the caterpillars of the brimstone and the blue-spotted hairstreak eat the leaves – possibly in your garden. This rare native bush is extremely versatile: it can tolerate everything from sun to shade and can even cope with marshy, damp or waterlogged soils.

TIP: Although the alder buckthorn is undemanding and its flowers are small, it is far from being a plain Jane. There are extravagant, low-growing varieties with delicate, almost fern-like foliage. These can be planted in small gardens, as a hedge or individual bushes. If the soil is compacted in some places, the selections below provide excellent solutions.

Varieties
- Asplenifoli (see photo): small variety with narrow leaves, up to 2 m (6.5 ft) high and 1.5–2 m (4.9–6.5 ft) wide
- Fine Line: delicate foliage, 1.5–2.5 m (4.9–8.2 ft) high and about 40–70 cm (1.3–2.3 ft) wide

PLANT PROFILE

- Wild form grows 3–5 m (9.8–16.4 ft) high and 2–4 m (16.5–13 ft) wide
- Small white flowers in May and June, second flowering usually continues into autumn
- For sun or shade
- Undemanding with regard to soil, as long as it isn't too dry, so it is also suitable for moist and even compacted soils
- Red berries from July, which turn black as they ripen, poisonous
- Turns yellow in autumn
- Deciduous native bush

Ice plant and orpine *Sedum spectabile, S. telephium*

These perennials are a paradise for insects, with many species romping around the flat heads of pink flowers at the same time in late summer. Bees, bumblebees, scorpion flies and butterflies arrive in hordes. Burrowing wasps on the hunt for wild bees often look for victims here, so you are guaranteed an exciting insect experience. As the flowers fade, they turn a deep purple colour. The decorative seed heads can be left standing all winter to provide a source of food for birds. These perennials are long-lived and can be easily propagated by division or cuttings.

TIP: A mining hoverfly of the genus *Cheilosia* may make the lower leaves of these perennials look rather unsightly in May. You can collect up and dispose of the fallen leaves, together with the larvae, but the attack is usually bearable. In my garden the great tits have learned to pick the grubs out of the leaves and are delighted by the easily accessible baby food for their chicks.

Varieties
- *Sedum spectabile* Iceberg: white flowers
- *S. telephium* Herbstfreude: classic, strong and robust
- *S. telephium* Matrona: hybrid purplish-green leaves
- *S. telephium* Purple Emperor: similar to Matrona, but even darker foliage

PLANT PROFILE
- Height and spread 40–70 cm (1.3–2.3 ft)
- Pink flowers from August to September
- For sun or partial shade – doesn't flower in full shade
- For normal garden soils, even if dry in summer

Hardy fuchsia *magellanica*

I confess that fuchsias are not really among my favourite plants. However, this attractive, hardy species from the Andes with its slender red and blue flowers that continue for months on end inspires such ardent enthusiasm in honey bees and bumblebees that it has won me over. A further argument in favour of this small shrub is the sausage-shaped, edible fruits, which you certainly can't buy in any supermarket. As the plant dies back to the base in harsh winters, it remains manageable in size, so it can also be recommended for small gardens. It is equally suitable for a herbaceous bed or the edge of a shrubbery. The large caterpillars of the elephant hawk moth like to eat the foliage.

TIP: Young plants will be slightly frost sensitive, so it is best to plant them out in spring. Covering them with autumn leaves also helps. If the fuchsia dies back during the winter, remove all the dead growth in spring and it will then shoot again from the base like a perennial. After mild winters you will not necessarily need to cut off the overhanging twigs.

Varieties
- *Fuchsia magellanica* var. *alba*: white and pale lilac flowers
- *F. magellanica* var. *globosa*: very low-growing at 50 cm (1.6 ft) in height, with relatively large flowers
- *F. magellanica* var. *gracilis*: very slender crimson and purple flowers

PLANT PROFILE
- Height and spread 60–100 cm (2–3.25 ft)
- Red and blue flowers from July to October
- Reddish edible fruits
- For sun, partial shade or light shade
- Likes a rich soil with plenty of humus that doesn't get waterlogged
- Also suitable for large containers

Wallflower *Erysimum cheiri*

This Mediterranean plant is a treasure trove for insects. In a short time one warm April day, on a good-sized patch, I observed hairy-footed flower bees, common mourning bees (see photo), ashy mining bees, bumblebees, a hummingbird hawk moth and, best of all, an appropriately coloured brimstone butterfly on the yellow flowers. Some species, such as the garden carpet moth, also make use of the leaves to feed their caterpillars.

Wallflowers are usually biennial but, by warm walls or in dry spots below the guttering, they can grow quite old and become a subshrub. They tolerate drought well, so they can even grow in cracks in walls. You can quite often see them at a dizzying height on old castle walls, but a less distant position is preferable for the sake of the scent.

TIP: In my mother's garden ancient wallflowers have been growing for decades in a very dry bed up against the house wall that gets scarcely any rain because of the projecting roof. This combination of food-plant and loose soil is ideal for ground-dwelling bees.

PLANT PROFILE

- Height 20–60 cm (8 in –2 ft), spread 30 cm (12 in)
- Scented yellow, red, orange or brownish flowers from April to June
- Likes sunny positions; tolerates poor, dry garden soil but not waterlogging
- Rarely attacked by snails
- Evergreen
- Easily propagated by seed

Wild teasel *Dipsacus fullonum*

The wild teasel and the closely related fuller's teasel are among the giants of our native flora. The prickly flower heads start to blossom in a ring around the middle, and then one ring makes its way up and another down. Bumblebees, hoverflies, wild bees and honey bees circulate in crowds around these pink pathways. Butterflies are also attracted.

The leaves grow together around the stem to form a funnel that can collect rainwater, as long as the teasel does not topple over. This imposing plant is biennial and dies after flowering. You can sow the seeds in spring, but on suitable soil it will seed itself successfully.

TIP: Leave the withered stems standing over the winter. They not only look fantastic frosted but also seem to go to the heads of colourful birds; goldfinches, traditionally known as thistle finches, eat the seeds and can locate and return to high-yielding teasel sites for years.

PLANT PROFILE

- Height 1.5–2 m (4.9–6.5 ft), spread 80–100 cm (2.6–3.25 ft)
- Pink flowers from July to August
- For sun or partial shade
- Likes rich soil that may also be quite moist, though also tolerates drought
- Rosettes of evergreen leaves
- Biennial

Mountain fleece *Polygonum amplexicaule*

This colossal member of the knotweed family from the Himalayas is an impressive perennial with spikes of red flowers and large heart-shaped leaves. It flowers non-stop from high summer until the frost, as if it wants to set a world record, delighting gardeners and bees alike. Paper wasps and hoverflies are also among its customers. With this plant, you will search for pests, especially snails, in vain.

Varieties with white, pink and wine-red flowers are available, so even pastel-colour-loving gardeners can get hold of their favourite shade of mountain fleece.

TIP: A merger with wild teasel and hemp agrimony is superb. The members of this unbeatable trio are well matched in size and desired position but splendidly contrasting in the shape and colour of their flowers. When the leaves and fruits of the spindle tree turn red in autumn, the mountain fleece will still be flowering and looking good next to Michaelmas daisies.

Varieties
- Album: white flowers
- Atropurpureum: ruby red flowers
- Blackfield: deep red flowers, leaves turn red in autumn
- Firetail: salmon pink, very abundant flowers and long flowering period
- Roseum: pink flowers

PLANT PROFILE
- Height 60–140 cm (2–4.6 ft), spread 80 cm (2.6 ft)
- Red flowers from July or August to October
- Perennial
- For sun or partial shade
- Likes rich, loamy, not-too-dry soil, but is extremely flexible
- Avoided by snails

Common poppy *Papaver rhoeas*

It is the epitome of summer: bright red with petals like silk. With poppies, you can give buff-tailed bumblebees in particular – but also honey bees and wild bees – a great deal of pleasure. The bumblebees are not always certain in their handling of the fluttering blooms and sometimes approach the black spots at the base of the petals from the wrong side. The seed capsules, which are shaped like salt pots, remain decorative for a long time.

The annual or biennial plant needs open soil to germinate and doesn't like competition. The seeds may wait in the soil for decades until the earth is dug over and they come into the light, which means that the enchanting display of red can be maintained by breaking up the soil or sowing again in March.

TIP: If sowing again every year is too much trouble for you, there is an alternative that is equally popular with insects from the realms of perennials: the oriental poppy (*Papaver orientale*), which has a similar flowering period. It is also available in more restrained shades. Its foliage dies down after flowering and doesn't appear again until autumn, making it the ideal plant if you go away in the summer.

PLANT PROFILE

- Height 20–90 cm (8 in–3 ft), spread 20–30 cm (8–12 in)
- Red flowers from May to July
- Annual or biennial – if the ground is dug over in autumn, the plants grow during the winter
- For sun to partial shade
- Likes well-drained soil that may be poor
- Copes well with drought

Globe thistle Echinops sphaerocephalus

The garden becomes a ballroom with the spherical flower heads of globe thistles above their prickly green leaves. Bees, bumblebees, butterflies and hoverflies squabble over the best places as they crawl around them. The globes continue to look attractive for a long time even after the flowers are over, and it is not necessary to cut them back since the plants will not usually flower a second time. This perennial is very drought tolerant and isn't bothered by hot summers. It works well at the foot of dry stone walls or in a gravel bed.

The globe thistle originates in Southern and Eastern Europe and has become naturalised in some places in the UK.

TIP: Globe thistles combine well with other drought-loving perennials. Achilleas create interest with their flat flower heads. Valerian and the native blue eryngo (*Eryngium planum*) can also be equal partners.

Varieties and other species
- *Eryngium banaticus* Taplow Blue: bright blue flowers, up to 1.2 m (4 ft) high, free-flowering
- *E. sphaerocephalus* Arctic Glow (see photo): white flowers from July to August, up to 1 m (3.25 ft) high
- *Eryngium* Veitch's Blue: not prickly; lower-growing than the species, 50–80 cm (1.6–2.6 ft) high; steel-blue flowers

PLANT PROFILE
- Height 60–150 cm (2–4.9 ft), spread 60 cm (2 ft)
- Blue flowers from July to August
- For sunny or part-shaded beds
- For poor dry soils that may also be stony
- Also tolerates normal, rich garden soil, as long as it doesn't get waterlogged
- Likes the heat

Lavender *Lavandula species*

Everybody loves lavender. Its scent is enchanting and evocative – there is no better plant to bring back memories of a Mediterranean summer holiday. Insects are also enraptured by the sea of blue flowers, so the evergreen bushes always host a large assembly of busy bees, bumblebees and butterflies.

As long as lavender is sited in well-drained soil it will withstand even harsh winters and may grow really old. It is best to plant several kinds with different flowering periods, which will greatly extend their offer of hospitality for insects. The flowering season starts off with the English lavender (*Lavandula angustifolia*), followed by the French lavender (*L. intermedia*).

TIP: Cutting back lavender immediately after flowering, including the uppermost leaves, and again down to about half in spring, will ensure that the plants don't become bare at the bottom. Any trimmings with flowers on them can be dried and used for aromatic lavender bags or scented cushions.

Species and varieties
- *Lavandula angustifolia* Hidcote Blue: dark-blue flowers, compact and vigorous, up to 40 cm (16 in) high
- *L. angustifolia* Two flowering seasons: June and again in September, 40–50 cm (1.3–1.6 ft) high
- *L. intermedia* Fragrant Memories: flowers from July to August, up to 80 cm (2.6 ft) high, strong scent
- *L. intermedia* Hidcote Giant: flowers from July to August, up to 60 cm (2 ft) high; especially good for bees

PLANT PROFILE
- 40–80 cm (1.3–2.6 ft) high and 50–80 cm (1.6–2.6 ft) wide, depending on variety
- Blue flowers from June to July or July to August, depending on species
- Needle-like, grey-green, highly aromatic foliage
- Likes a sunny position
- For poor dry soils, also on walls
- Avoided by snails

Purple tansy *Phacelia tanacetifolia*

You could consider this American annual simply as a green manure, since it loosens up the soil a long way down and delights the earthworms. However, you can also sow it quite selflessly to create a pretty blue bee meadow that will flower all through the summer, providing plenty of nectar and pollen. Young and old, everybody loves the purple tansy, sometimes called the bee plant – from bumblebees and honey bees to the tiny sweat bee that just clings to the long stamens with their blue pollen. Hoverflies do the same, until their stomachs are full and the blue colour shines through from the inside. You will have colourful summer beds if you sow common poppies at the same time.

TIP: After flowering, I don't cut the plants down to ground level until the spring, and I leave digesting the roots to the earthworms. Then it's a matter of self-seeding. The seedlings usually survive the winter even in the event of hard frost, get an early start and quickly come into flower. Sow them in spring in spots that are empty, to extend the flowering period.

PLANT PROFILE

- Height 30–120 cm (12 in–4ft), spread 20 cm (8 in)
- Blue flowers from June to September
- Annual, also biennial if self-sown in autumn and overwintered
- For sun to partial shade
- Likes garden soil that doesn't get waterlogged; can cope with few nutrients and with drought
- Easy to grow from seed; sow directly in the bed from March to July
- Self-seeding
- Green manure

Scots rose *Rosa spinosissima*

This rose always wants to be first and usually flowers in May before all the other native wild roses. What's more, it is visited by many different insects: bumblebees, hoverflies, mining bees and honey bees – as well as a large number of smallish beetles – are attracted by the pure white flowers with yellow stamens. It is a robust and drought-tolerant shrub. Unusual features are the black hips that ripen early and the particularly delicate leaves with lovely autumn colour. The prickly suckers are less of a virtue.

TIP: The delicate shape of this rose is deceptive; underground, it strives for world domination, putting out numerous suckers. This is ideal for stabilising a slope, but not in a flowerbed. Either use improved varieties or plant unimproved bushes where they will be restricted by paving or a root barrier. As not all wild forms flower equally well, have a look at the bushes in flower and propagate good specimens by layering.

Varieties
- Frühlingsgold: yellow flowers in June; no suckers as it is an improved variety
- Red Nelly/Single Red: 1.5 m (4.9 ft) high, deep violet flowers; no suckers as it is an improved variety
- *Rosa spinosissima* var. *lutea*: 90 cm (2.9 ft) wide and high with yellow flowers

PLANT PROFILE

- 0.5–2m (1.6–6.5 ft) high and 1.5 m (4.9 ft) wide, depending on position
- White, scented flowers in May; sometimes a small second flowering in late summer
- Very small leaves, which are colourful in autumn
- Many downward-pointing thorns, which are red on young growth
- Black hips
- Likes sunny or semi-shady positions
- For thin, dry sandy soils, but also for normal garden soils
- Produces many suckers

Many-flowered rose *Rosa multiflora*

This Asian wild rose is found in many gardens, but mostly incognito as a decorative under layer. There, it is good to restrict it to small areas as it grows into an enormous bush, which can only be kept under control to some extent by thinning out after flowering. During its exuberant flowering period all its delusions of grandeur will be forgiven. Honey bees, bumblebees and quite a few wild bees love the small white flowers so much that the bush will be buzzing, especially in the mornings. For this reason, hornets like to use this rose as a hunting ground, and catch the occasional honey bee there. Songbirds like to eat the tiny hips, which are arranged in clusters like the flowers.

TIP: Many rambler roses are descended from the many-flowered rose, although the latter does not climb well. If you want to breathe new life into an old or completely dead tree, multiflora ramblers such as the snow-white Rambling Rector or the violet Veilchenblau are good as their flowers are not too full of petals.

Variety
- Nana: small bush rose, up to 1 m (3.25 ft) high

PLANT PROFILE
- Height and spread 3 m (9.8 ft), tolerates cutting back
- Small white flowers with yellow stamens in bunches from June to July, scented
- Pea-sized red hips
- Few thorns
- For sun or partial shade
- Likes any humus-rich, well-drained soil
- Self-seeding
- Does not produce suckers
- Easy to grow from cuttings

Bee tree   *Tetradium daniellii*

It is a real shame that our native flora has not produced anything as fantastic as the bee tree, also known as Korean evodia. Instead, the fame and glory goes to Asia, where this shrub has its home. It is hardy here, too, however, and can be used very successfully to enrich our gardens as a small domestic tree, to the delight of numerous insects.

With its luxuriant white blossom, the bee tree fits perfectly into the summer gap, when our native trees are taking a break from flowering. Bumblebees, honey bees, butterflies, flies, wasps and even small wild bees gratefully accept the offer and harvest nectar and pollen. Later on, the fruits are eaten by birds.

TIP: Young plants can easily be grown from seed, but they will need to be protected from hard frosts during their first two winters. However, as the seedlings only begin to flower after about five years, a bought tree will give you a big lead.

PLANT PROFILE

- Height 4–8 m (13.1–26.2 ft) (rarely 10 m/32.8 ft), spread 3–5 m (9.8–16.4 ft)
- Large panicles of white flowers from July to August, often flowering a second time, up until September
- Bright red seed heads from September
- Shiny black seeds
- For a sunny or part-shaded position
- Likes rich soil that can be loamy or sandy
- Spring planting is recommended

Hollyhock *Alcea rosea*

Hollyhocks are as much a part of summer as ice lollies. These towering members of the mallow family still have a place everywhere, despite their size. Gigantic hollyhocks can even sprout from cracks in a sunny house wall. They are large and classy and can produce more than 50 flowers per stem. Their big leaves make it almost incredible, but thanks to their taproots these short-lived perennials are very drought tolerant, which helps them even more in extreme positions such as cracks in pavements. Honey bees like to visit the big flowers, while bumblebees are obliged to fly round in circles like aircraft in a holding stack.

The triangular nectar sources are at the base of the flower. Insects cover themselves with layer on layer of pollen as they dine. A few fun-loving weevil species, such as the hollyhock weevil, eat their way through the buds.

TIP: If the position is not sufficiently sunny and airy, hollyhocks will suffer from hollyhock rust, which damages the leaves. If the plants in your garden are not doing well, just try out the new hybrids of hollyhock and marsh-mallow – for example the Parkallee variety – which are less susceptible to the fungus.

PLANT PROFILE

- Height 1–2.5 m (3.25–8.2 ft), spread 50 cm (1.6 ft)
- White, yellow, pink, red or black flowers from July to September
- For sunny to part-shaded positions
- Likes rich soils, copes well with drought
- Sometimes only biennial but more usually perennial
- Flowers for the first time in the second year, often more than one flower stem from the third year on
- Seeding works well

Cranesbill Rozanne *Geranium hybrids*

So blue for so long – only a few geraniums can do that. In good years, the Rozanne variety – previously also available under the descriptive name Jolly Bee as they were thought to be different varieties – flowers all the way through from May to October. It can afford to do so because it is sterile and does not need to invest a lot of energy in seed production.

The blue flowers with pale centres are large and grow on long, flexible stems, so that a plant could do with 1 sq m (3.25 sq ft) of bed all to itself and will grow through other plants if need be. Its customers include honey bees, butterflies and many small wild bees, including yellow-faced bees and sweat bees. Male solitary mason bees are very fond of this variety, too. When combined with bellflowers, they make a total package for this species of wild bee.

TIP: Don't worry – although this variety is late in sprouting it always gets back into shape. Its pendulous growth makes this perennial suitable for large balcony containers and hanging baskets. In late summer Rozanne remembers its core business and produces new shoots from below, allowing you to remove the old tentacle-like flower stems.

PLANT PROFILE

- Height 30–60 cm (12–24 in), spread 1–1.5 m (3.25–4.9 ft)
- Blue flowers from May until frost
- For sunny to part-shaded positions
- Likes humus-rich, not-too-dry soils
- Grateful for added compost
- Not threatened by snails

Hemp agrimony *Eupatorium cannabinum*

Hemp agrimony is not only one of our most imposing native perennials but it's also great for insects of all kinds. Bees, bumblebees and hoverflies like to land on the masses of pink flowers that appear throughout the summer. Butterflies also find it excellent and red admirals, peacocks and small tortoiseshells in particular are likely to be seen on it. It is also a food-plant for the caterpillars of many species of butterfly and moth, making it an ideal occupant for a garden that is close to nature. As this species likes damp soils, it is at home on pond edges, but it will also grow in any loamy soil that retains moisture.

TIP: Hemp agrimony spreads widely and happily seeds itself, meaning that it is recommended more for larger gardens. However, as it flowers for months on end, it also enriches smaller gardens – then you can simply plant it instead of a shrub, including for screening.

PLANT PROFILE

- Height 1–1.5 m (3.25–4.9 ft), spread 1 m (3.25 ft)
- Pink flowers from July to September, the intensity of the colour varies
- Perennial
- Likes sun or partial shade
- For rich, moist soils
- Self-seeding

Almond willow *Salix triandra* Semperflorens

Willows are very valuable because they flower early and always provide a welcome meal for the first bees of the season. However, this variety, which is native to Europe, simply forgets that spring comes to an end at some point and keeps on flowering until the autumn. Its main flowering period is actually in summer, when many other bushes have already finished flowering. This species of willow is therefore a reliable resort for honey bees and wild bees, and it is also a good food-plant for the caterpillars of moths, such as the red-tipped clearwing, which eat into the wood.

Male plants, which have yellow catkins with plenty of pollen, are the most commonly sold. The bush is happy on any soil, but prefers to have its feet wet and is often found growing on river banks.

TIP: If not cut back, this willow will grow quite large, but it can be pollarded. To do this, cut it back hard in the spring. The twigs can be woven into fencing or used to support perennials.

PLANT PROFILE

- Height 2–6 m (6.5–19.7 ft), spread 2–4 m (6.5–13.1 ft), tolerates some cutting back
- Very slender flowers from April to October, male flowers have yellow pollen, female flowers green
- Deciduous bush or small tree with narrow leaves
- For sun and partial shade
- Suitable for damp soils, but can also cope with dry positions
- Shallow rooted
- May be propagated by cuttings
- Turns yellow in autumn

Filler plants
Another helping

They will always fill a gap somewhere, even in shady spots. There may not be great crowds around these niche products, but mining bees and other individualists will find a meal here, away from the hurly-burly.

Wild garlic *Allium ursinum*

Wild garlic is a plant for relaxed gardeners. It feels really at home where the autumn leaves are allowed to rot under the bushes and the ground is not dug over. It is therefore less suitable for newly laid-out beds than for old ones that have been able to work on creating a humus-rich soil for several years.

PLANT PROFILE

- Height 20–40 cm (8–16 in), spread 20 cm (8 in)
- White flowers from April to May
- Perennial that dies down after flowering
- For partial shade or shade
- Likes rich soils with a good layer of humus
- For around shrubs and under deciduous trees
- Self-seeding

This small perennial with globes of white flowers allows all its shiny foliage to die down in early summer, thus successfully avoiding the great heat of summer. If only for its culinary aspects, wild garlic is a must in shady parts of the garden, where it will provide good ground cover on a part-time basis. Among the bees it is mostly small wild bees that visit the flowers. In my garden it mainly attracts mining bees, but bee flies also come along (see photo).

TIP: Buy potted plants or ask friends to give you offshoots, as wild garlic can get a bit uptight when you try to grow it from seed. Not only is it a cold germinator, but it is also a real diva and can take several years to germinate. However, once it is planted out in the garden it overcomes its shyness and suddenly pops up in every possible, and impossible, corner.

Leopard's bane *Doronicum pardalianches*

Leopard's banes have always had a regular place in the spring garden, yet this species is quite rarely found. It is difficult to plant in a flowerbed, as it produces runners (though these do not become a nuisance) and doesn't like to be kept in place. It continually changes its position, which is actually an advantage in areas of light shade around shrubs, where its small, sunflower-like blooms can bring life under the bushes.

What is especially handy about this perennial is its habit of dying down completely after its many feathered seeds have flown the nest in summer, so you don't have to bother about it in summer drought. The heart-shaped leaves always reappear, as fresh as ever, on the surface in autumn and last throughout the winter. Wild bees like to make their way to the yellow flowers.

TIP: This perennial goes well with other fickle woodland-edge plants, such as wood forget-me-not, celandines and columbines. Solomon's seal, wild garlic and great forget-me-not make suitable contrast plants.

Variety
- Doronicum Goldstrauss: many-branched flower heads

PLANT PROFILE

- Height 50–80 cm (1.6–2.6 ft), spread 40 cm (16 in)
- Yellow flowers from May to July
- Perennial
- For partial shade or light shade
- Likes rich soils, drought tolerant
- Perfect for the edge of shrubs
- Produces runners
- Also suitable for tubs and boxes

Ground ivy *Glechoma hederacea*

If I can hear buzzing somewhere close to my feet, the ground ivy is not far away – with bees, hoverflies and bumblebees always lured by its charms. Hairy-footed flower-bees in particular are extremely fond of it. The blue flowers of this native wild perennial stretch upwards on vertical stems, but otherwise the plant creeps horizontally close to the ground, attracting insects underneath the bushes. Butterflies such as cabbage whites, brimstones and orange tips suck nectar from the flowers, and the caterpillars of a number of butterflies and moths feed on the leaves.

Ground ivy isn't much of a bother for the gardener, being a ground-cover plant that needs little attention.

TIP: Ground ivy has fallen out of favour with many gardeners, who look on it as a weed because it can quickly cover large areas with its inquisitive tendrils. However, it not only keeps the ground moist but is also a tasty wild herb for salads and vegetable dishes. Having once tried it, I now welcome every new shoot as a rare and special delicacy.

Variety
- Variegata: green and white variegated leaves; not always hardy, keep some cuttings in the house over the winter to be on the safe side

PLANT PROFILE

- Height 5 cm (2 in), up to 20 cm (8 in) including flowers, ground-cover plant
- Blue flowers from March to May
- Perennial with aromatic leaves
- For sun, partial shade or light shade
- Likes rich, not-too-dry soils
- Grows happily under bushes
- Produces runners
- For cooking with wild herbs

Bluebells *Hyacinthoides*

In spring, bluebells transform the English woodlands into a sea of blue flowers. In this case it is the native common bluebell (*Hyacinthoides non-scripta*). Anyone who has seen photos of it immediately wants to have a similar waving meadow of flowers in the garden, even if on a more modest scale. The Spanish bluebell (*H. hispanica*) and hybrids of the two species (*H.* × *massartiana*) are more usually available to buy. The Spanish bluebell is also hardy.

If you want to contribute to the conservation of the native bluebell, seek out suppliers who propagate this species. These bulbs find a place anywhere under deciduous shrubs where there is a layer of compost, grow wild over time and attract wild bees in particular. In my garden, these are mostly mining bees.

TIP: In our local park there is a secret place with a fantastic trio of unknown origin: bluebells growing together with early crocus (*Crocus tommasinianus*) and glory of the snow (*Chionodoxa*) under maple trees. These plants also get on well with each other in the garden and provide flowers from March to May.

PLANT PROFILE

- Height 20–30 cm (8–12 in), spread 20 cm (8 in)
- Blue, pink or white flowers from April to May
- Bulbous plant
- For partial shade or shade, happy with more light in spring
- Likes humus-rich soils that can be moist but not waterlogged
- For the woodland edges and under deciduous trees and shrubs
- Planting time for the bulbs is September to November
- The leaves should be left to turn yellow

Ornamental onion *Allium aflatunense*

This large onion is a clever filler for a sunny spot. It puts on a great show as soon as the flowers begin to open. The violet globes look good with aquilegias. Bumblebees, honey bees, mining bees and other insects love to gather round this onion. Ruby-tailed wasps, which parasitise the nests of wild bees in the insect hotel, often go for the centre of the flower head for a place to sleep.

As the seeds ripen, a green ball is formed, which later turns brown and ensures that there will be many seedlings. So don't cut off the dead heads; they will continue to look decorative for a long time. Bulbils sometimes develop at the base of the flowers, and these can be planted. If dried, the hollow stems can have a second career in the bee hotel.

TIP: Before I became well acquainted with the ornamental onion, I was really frightened when I saw that the leaves were already turning yellow during flowering. Not even an extra portion of water stopped the withering. Now I know that it is not suffering from lack of water but that it simply doesn't need the leaves any more; it has already stored enough food in its bulb and finishes flowering and seed-ripening on its reserves.

Varieties and other species:
- Portuguese allium (*Allium lusitanicum*): pink flowers from July to September; for stony, extremely dry positions
- Purple Sensation: large violet flowers on long stems
- Round-headed leek (*A. sphaerocephalon*): lilac flowers in July; very slender native species, up to 1 m (3.25 ft) high

PLANT PROFILE
- Height 50–90 cm (1.6–3 ft), spread up to 20 cm (8 in)
- Lilac flowers from May to June
- Perennial bulbous plant that dies down in May
- For sun or partial shade
- Undemanding with regard to soil, as long as it doesn't get waterlogged
- Tolerates drought
- Sows itself well, germinates in spring
- The bulbs are planted in autumn

Spring snowflake *Leucojum vernum*

The spring snowflake often flowers in February, at the same time as the snowdrop. It bravely fights its way through the covering of snow and is not deterred, even by night frosts. You can combine it with snowdrops, but it is not quite as undemanding and eager to reproduce as its relative. It takes time and patience to build up a big stock. However, it is worth the wait, as bees and other insects like to come along when the beautiful white flowers with the green tips are dancing above the snowdrops.

TIP: Like snowdrops, spring snowflakes are best propagated with the leaves on – that is, in spring, after flowering. The plants mustn't dry out on their way into your garden. Propagation by seed is also possible, but it is not for the impatient as the seedlings need several years before they'll flower for the first time. For a random spread, just place ripe seed capsules near an ant nest and let the six-legged customers do the work.

PLANT PROFILE

- Height 10–30 cm (4–12 in), spread up to 20 cm (8 in)
- White and green flowers between February and March
- Perennial native bulbous plant
- For meadows, pond margins, under deciduous shrubs and along the undersides of hedgerows
- Likes a rich, loamy soil that may be moist and should be rich in humus
- The seeds are spread by ants

Agrimony *Agrimonia procera*

You have probably seen the common agrimony (*Agrimonia eupatoria*) flowering in meadows. Its big brother is less well known, but is in its proper place in gardens as there its impressive appearance is not wasted and is hard to overlook. It can also tolerate more shade than its relatives. The yellow flowers, which are visited by bees and hoverflies, are arranged close together on long spikes with a few forks. The leaves give off a scent when touched. When the seeds ripen, the plant tends to brush up against the gardener because its fruits are little burrs that particularly like to catch on woolly jumpers and thus get all over the garden. The seeds germinate easily.

TIP: The plant produces a strong root, which makes life easier for it in dry places but makes transplanting difficult. However, the seedlings are easily identified by the striking scent of the leaves, so you can move them to suitable positions in good time. The plant looks especially lovely in small groups, lending the flower-spikes impact even at a distance.

PLANT PROFILE

- Height 50–180 cm (1.6–5.9 ft), spread 20 cm (8 in), slender, upright growth
- Native perennial with yellow flowers from July to August
- Aromatic leaves
- For sun, partial shade or light shade
- Undemanding, tolerates poor, dry soil
- Likes to grow beside hedges
- Self-seeding
- Avoided by slugs

Snowdrop *Galanthus nivalis*

Fortunately, snowdrops are found in many gardens, but you often have to be a keen-eyed observer to spot insects on them. If the overall supply of flowers is still limited on warm days in early spring, flies, honey bees, mining bees and sometimes even butterflies will look in. You don't need any expensive collectors' items for this; the visitors' favourite is the common or garden snowdrop. The small bulbous plants like to grow under shrubs and trees, and in areas of sparse grass. After they have produced their seeds they disappear back underground, so they are also well suited for integration in shady corners.

TIP: Bulbs offered for sale in the autumn don't usually grow because they dry out easily. Propagation on the principle of divide and rule is more successful.

Clumps of snowdrops can be dug up and divided in two after flowering. Clumps that get too big will stop flowering anyway, so people are usually more than happy to give them away as gifts to other gardeners after dividing them.

PLANT PROFILE

- Height 10–20 cm (4–8 in), spread up to 20cm (8 in)
- White and green flowers between February and March
- Bulbous perennial; the grey-green leaves disappear in early summer
- For areas under bushes where there is summer shade
- Likes chalky, rich, not-too-dry soil
- The seeds are spread by ants

Greater celandine *Chelidonium majus*

The native celandine is not exactly popular with gardeners as it seeds itself everywhere, and it makes a negative impression when weeding because its milky, orange sap dyes the skin. However, it must be admitted that it will take over areas that are difficult to green, such as shaded spots or crevices in walls. Even being close to large conifers cannot restrain its enthusiasm. I always leave one or two of these distinctive plants standing at the back of the garden, much to the delight of the bumblebees and honey bees, who appreciate its long flowering period. In addition, a few moths make use of the foliage for laying their eggs.

TIP: Very small seedlings can be confused with aquilegias, which are similar in shape to the greater celandine. The latter are characterised by the grey-green colour of the leaves and the emission of the milky, orange sap. You soon develop an eye for the seedlings and can weed them out if you have too many – and your fingers won't be dyed such a strong colour as with older plants.

PLANT PROFILE

- Height 20–80 cm (8 in–2.6 ft), spread up to 40 cm (16 in)
- Yellow flowers from May to October
- Short-lived perennial
- For sun or shade
- For any soil; even poor, dry or sandy soils will be populated
- The seeds are spread by ants
- Medicinal plant; the sap was once used against warts

Knotted cranesbill *Geranium nodosum*

This delicate cranesbill from the mountains of Europe is phenomenal: it has lovely shiny leaves, flowers without interruption in the shadiest spots and is really drought tolerant, even under bushes. The ripening fruits with red tips are pretty as well.

This perennial can be planted under hedges and deciduous shrubs, where its pink flowers shine out of the darkness. Bumblebees, hoverflies, mining bees and sweat bees (see photo) like to visit the flowers. Mice nibble the schizocarpic fruits while they are green.

TIP: Like all cranesbills, this species is a master of the long throw. Its catapult mechanism flings the seeds several metres away from the plant. Of course, it can't take aim, so not all of them hit the ground. However, wherever they land successfully the seeds start to germinate, even in a tomato container. This perennial can be recognised by its shiny leaves, even in early youth. As a result, I have been able to give away plenty of small plants, which is handy since this species can't be bought in garden centres.

Varieties
- Clos du Coudray: darker flowers with a white border
- Silverwood: white flowers
- Simon: flowers larger and more abundant than the species
- Svelte Lilac: more intense colour than the species, but not so vigorous

PLANT PROFILE

- Height 40–50 cm (1.3–1.6 ft), spread up to 50 cm (1.6 ft)
- Pink flowers with lilac stripes from May to October
- Perennial
- For partial shade or shade
- Likes a well-drained position with plenty of humus; grows on acid soils
- Spurned by slugs

Wood forget-me-not *Myosotis sylvatica*

Without the native forget-me-not my newly laid-out flowerbeds would have looked very bare. This plant, which is usually biennial, overwintering as a rosette of leaves flowering in the second year, is always ready to leap into the breach and quickly fills the gaps. The plants, which still look delicate in March, absolutely explode and quickly form bushy flower heads that work at full stretch for months on end. Honey bees, mining bees and mason bees find them a reliable source of nectar that is also used by butterflies. The shiny black seeds are eaten by bullfinches and other finches.

As the sticky seed cases are very clinging and get caught on clothing, they travel quite a long way around the garden. They also seed themselves abundantly, as well as being spread by ants – so there is no risk of ever having too few sky-blue flowers in the garden.

TIP: If the plants start to look unsightly in June with only the odd individual flower, I remove them and shake them out over all the beds. Next, I throw them on the compost heap, from where the remaining seeds later start off on their travels.

PLANT PROFILE

- Height 15–30 cm (6–12 in), spread up to 40 cm (16 in)
- Pale blue, occasionally white, flowers with yellow centres from April to June
- For sun or partial shade
- For any soil that isn't waterlogged
- Abundantly self-seeding
- Also suitable for pots and window boxes
- No risk of being eaten by snails

Hedge woundwort *Stachys sylvatica*

If the flowers were just a little bit bigger, this native perennial would certainly have a great career ahead of it; nevertheless, it will grow anywhere in the garden, although it is inclined to choose hedgerows and under bushes. Where it is allowed to, hedge woundwort also wanders through herbaceous beds. Surplus plants can easily be pulled out.

Bumblebees love the purple flowers with their unusual white markings. If there is a good stock of them in the garden, plus a good measure of dead wood nearby, hairy solitary bees (*Anthophora furcata*, see photo), whose favourite plant is hedge woundwort, will fly in. Still not convinced? Then look at the hairy flower heads with the light behind them – they simply look magnificent.

TIP: Although the leaves smell a bit disconcerting, use them raw in a salad. They taste of lemon to begin with, then like raw mushrooms.

PLANT PROFILE

- Height 30–120 cm (1–4 ft), spread 20 cm (8 in)
- Lilac flowers with white flecks from June to August
- Strong-smelling leaves with white hairs, with a shape reminiscent of stinging nettles
- For sun or shade
- Likes rich, not-too-dry positions under bushes and in hedges
- Produces runners and is inclined to self-seed
- Also grows in containers

PERENNIALS FOR BUMBLEBEES

English name	Botanical name	Flowering period	Characteristics
Big betony	Stachys grandiflora	VII–VIII	For sun or semi-shade
Blue comfrey	Symphytum azureum	IV–V	Sky-blue flowers
Bugle	Ajuga reptans	IV–VI	Native perennial for shade, invasive
Candle larkspur	Delphinium elatum	VI–VII	Watch out for snails
Common comfrey	Symphytum officinale	V–VII	Native, for damp soils
Common milkweed	Asclepias syriaca	VI–VIII	From North America, parrot-like fruits
Crocus	Crocus	II–III	To grow wild in scruffy lawns
Dusky cranesbill	Geranium phaeum	V–VII	Self-seeding. Some varieties with spotted leaves
English name	Botanical name	Flowering period	Characteristics
Eryngium	Eryngium	VII–IX	Bizarre flowers
Goat's rue	Galega hartlandii Alba	VI–IX	Sterile variety, not self-seeding
Grape hyacinth	Muscari	IV	Self-seeding
Iranian wood sage	Teucrium hircanicum	VI–X	For dry areas from sun to light shade. Self-seeding
Japanese anemone	Anemone hupehensis	VII–X, depending on variety	Spreads quickly with runners
Large-leaved lupin	Lupinus polyphyllus	VI–VIII	Many varieties
Lesser calamint	Clinopodium nepeta	VII–IX	Continuously flowering, for sunny spots
Meadow sage	Salvia pratensis	VI–VIII, IX	Native perennial for low-fertility meadows
Plume thistle Atropurpureum	Cirsium rivulare	VII–VIII	Sterile variety of the native species
Solomon's seal	Polygonatum multiflorum	V–VI	Blue berries for the birds
Spotted dead-nettle	Lamium maculatum	V–VII	Many varieties of leaf decoration
Spring pea	Lathyrus vernus	IV–V	Native woodland-edge perennial
Sticky sage	Salvia glutinosa	VII–IX	Yellow flowers, native woodland-edge perennial
Stinking hellebore	Helleborus foetidus	II–IV	Evergreen, native species for semi-shade
Thyme	Thymus vulgaris	VII–IX	For dry areas
White dead-nettle	Lamium album	IV–X	Long-flowering in shade
Whorled sage	Salvia verticillata	VI+IX	Will flower again after cutting back
Woodland sage	Salvia nemorosa	VI–IX	Second flowering after cutting back

SHRUBS FOR BUMBLEBEES

English name	Botanical name	Flowering period	Characteristics
Barberry	Berberis vulgaris	IV–VI	Edible berries, native
Butterfly bush	Buddleia davidii	VII–IX	Also interesting for butterflies
Common broom	Cytisus scoparius	V–VII	Varieties in various colours, native
Flowering currant	Ribes sanguineum	IV–V	Edible berries
Fly honeysuckle	Lonicera xylosteum	V–VI	Red berries for birds
Goat willow	Salix caprea	III	Early willow for bees, tolerates cutting back
Mezereon	Daphne mezereum	III–IV	Native dwarf shrub, poisonous
Oregon grape	Mahonia aquifolium	III–VI	From North America, edible blue berries
Purple broom	Chamaecytisus purpureus	V–VI	Small shrub for dry walls
Rock rose	Helianthemum	V–VII	For the rock garden, dwarf shrub
Rose of Sharon	Hibiscus syriacus	VII–IX	Many varieties
Seven son flower tree	Heptacodium miconioides	VI, VIII–XI	From China, long flowering period in poor summer gap
Snowberry	Symphoricarpos	VI–VIII	Varieties with white and red berries for birds
Snowy mespalis	Amelanchier ovalis	IV–V	Edible berries
Spiraea bumalda	Spiraea x bumalda	VII–IX	Long flowering period
Trailing broom	Chamaecytisus supinus	V–VIII	Yellow flowers and compact for dry soils
Tutsan	Hypericum androsaemum	VI–VIII	Small shrub with large yellow berries

BEE-FRIENDLY CLIMBERS

English name	Botanical name	Flowering period	Characteristics
Alpine clematis	Clematis alpina	IV–V	Also suitable for large containers
Broad-leaved everlasting pea	Lathyrus latifolius	VI–IX	Drought-tolerant perennial
Climbing hydrangea	Hydrangea petiolaris	VI–VII	Climbs with anchoring roots on walls or tree trunks
Common ivy	Hedera helix	IX–X	For sun or shade
Field rose	Rosa arvensis	VII	Native wild rose that can be trained to weave into fences
Virginia creeper	Parthenocissus	VI–VII	Climbs using anchoring roots or tendrils, depending on variety
Wisteria	Wisteria	V–VI	Strong-growing twining plant

PLANTS FOR SPECIALISED WILD BEES

English name	Botanical name	Characteristics	Bee species
Bellflower	Campanula	Many native perennials for different walks of life	Gold-tailed melitta (Melitta haemorrhoidalis), solitary mason bee (Osmia rapunculi)
Bryony	Bryonia	Native climbing perennial, seeds abundantly	Bryony mining bee (Andrena florea)
Field scabious	Knautia arvensis	Perennial for rich soils	Large scabious mining bee (Andrena hattorfiana)
Goat willow	Salix caprea	Cuttable shrub, pendulous cultivated form	Grey-backed mining bee (Andrena vaga)
Greater knapweed	Centaurea scabiosa	Perennial for poor, dry, sunny spots	Anthidium species
Honesty	Lunaria annua	Biennial, seeds itself	Osmia brevicornis
Mignonette	Reseda	Annual or biennial species	Large yellow-faced bee (Hylaeus signatus)
Round-headed leek	Allium sphaerocephalon	Native species for sunny spots	Onion yellow-faced bee (Hylaeus punctulatissimus)
Yellow loosestrife	Lysimachia vulgaris	Perennial for fairly moist soils	Macropis (genus)
Viper's bugloss	Echium vulgare	Biennial for poor soils in full sun	Viper's bugloss mason bee (Osmia adunca)

Index

Acanthus 56
  hungaricus 57
  mollis 57
Achillea 9, 33, 35, 42, 54, 68, 98
A. filipendulina 9, 34, 35, 42, 54, 68
A. millefolium 34
Aconitum
  carmichaelii 70
  napellus 65, 70
Agastache 75
A. rugosa 67
Agrimonia procera 116
Agrimony 116
Ajuga reptans 28, 122
Alder buckthorn 91
Allium
  aflatunense 77, 114
  lusitanicum 114
  sphaerocephalon 42, 114, 124
  ursinum 110
Anchusa 80
Andrena
  florea 124
  hattorfiana 124
  vaga 124
Anthemis tinctoria 34, 36, 54, 55, 78
Anthidium
  manicatum 22
  nanum 124
Anthophora plumipes 28
Apple 48, 63, 88
Aquilegia 10, 64, 66, 111, 118
Archangel, balm-leaved 79
Aubrieta deltoidea 28, 48, 56

Ballota nigra 22, 24
Balm
  bastard 76
  spotted bee 75
Balm-leaved archangel 79
Baptisia australis 44, 46
Bastard balm 76

Bear's breeches 56, 57
Bee tree 103
Bellflower 9, 11, 124
  clustered 25, 50, 51
  giant 53
  nettle-leaved 50
  peach-leaved 50
  rampion 50
  trailing 50, 52
Bergamot 75, 90
Betony 22, 25, 26
  big 122
Bilberry 20
Bladder-senna 44, 45
Blue eryngo 64, 98
Bluebell 58, 113
Bombus
  hortorum 65
  hypnorum 63
  pascuorum 62
  terrestris 64
Borage, early-flowering 28, 32, 33, 62
Box, common 20
Buckthorn, alder 91
Buddleia davidii 56, 123
Buff-tailed bumblebee 64, 69
Bugle 28, 122
Bugloss, common 80
Bumblebee
  buff-tailed 64, 69
  garden 65, 72
  tree 63
Buphthalmum speciosum 54
Buxus sempervirens 20

Calamint 38
  lesser 122
Campanula 124
C. glomerata 25, 50, 51
C. latifolia 50, 53
C. persicifolia 50
C. poscharskyana 50, 52
C. rapunculus 50
C. rotundifolia 50
C. trachelium 50

Carder bee
  common 62, 72
  wool 22, 24, 26, 27, 59, 71
Carpenter bee 12, 57, 58, 59
  violet 47, 56
Catmint 9, 68, 77
Celandine, greater 111, 118
Centaurea montana 40, 62, 72
Chamomile, yellow 34, 36, 54, 55, 78
Colletes
  bare-saddled colletes 34, 36
  C. hederae 38
  C. similis 34, 36
  Davies' 34, 36, 37, 54
Colutea arborescens 44, 45
Comfrey 10
  blue 122
  common 28, 65, 122
  creeping 30
Coneflower
  orange 90
  purple 64, 68
Cornflower, perennial 9, 40, 62, 72
Corydalis 10
  C. cava 28, 31, 64
  C. solida 28, 31
Cranesbill 10, 40, 42, 50
  dusky 122
  knotted 119
  meadow 9
  stinking 17
Crocus 11, 16, 48, 64, 113, 122
Culver's root 74, 90
Currant 20, 21

Davies' colletes 34, 36, 37, 54
Dead-nettle 10, 28
  giant 79
  spotted 62, 79, 122
  stiff 22
  white 62, 79, 122
Digitalis
  grandiflora 71

lanata 22, 71
lutea 65, 71
purpurea 22, 71
Dipsacus fullonum 62, 95
Doronicum pardalianches 111

Echinacea purpurea 64, 68
Echinops
  banaticus 98
  sphaerocephalus 98
Echium
  plantagineum 9
  vulgare 9, 78, 124
Elecampane 54, 55
Eryngium 122
E. planum 64, 98
Eryngo 64, 98
Erysimum 48
E. cheiri 28, 94
Eupatorium cannabinum 106
European orchard bee 48
Everlasting pea, broad-leaved 44, 47, 56, 124

False indigo 44, 46
Feverfew 34
Forget-me-not, wood 111, 120
Fork-tailed flower bee 12
Foxglove 10
  common 71
  large yellow 71
  small yellow 65, 71
  woolly 22, 71
Frangula alnus 91
Fuchsia magellanica 93
Fuchsia, hardy 93
Fumewort 28, 31

Galanthus nivalis 117
Garden bumblebee 73
Garlic, wild 110, 111
Geranium 40, 42
G. nodosum 119
G. phaeum 122
  Rozanne 105
Germander
  Chinese 58
  Iranian wood sage 122
  wall 22

Glechoma hederacea 28, 112
Globe thistle 98
Gold-tailed melitta 9, 124
Ground ivy 28, 112

Hairy solitary bee 121
Hairy-footed flower bee 28, 94, 112
Halictus 40
Harebell 50
Hedera helix 38, 39, 124
Hellebore, stinking 76, 122
Hemp agrimony 96, 106
Herb-robert 17
Holewort 28, 31, 64
Hollyhock 104
Honeysuckle 65, 73
Hyacinthoides 113
Hydrangea, climbing 124
Hylaeus 42
  communis 42
  nigritus 9, 37, 42
  punctulatissimus 12, 42, 124
  signatus 42, 124
Hyssop, giant 75
Hyssopus officinalis 82

Ice plant 35, 62, 92
Indigo, false 44, 46
Inula 54
  helenium 55
Ivy
  common 38, 39, 124
  ground 28, 112
Ivy bee 38

Knapweed, greater 124
Korean mint 67

Lamb's ear 22, 27, 44, 56, 68
Lamium 28
L. album 62, 79, 122
L. maculatum 79, 122
L. orvala 79
Lasioglossum 40
Lathyrus
  latifolius 44, 47, 56, 124
  odoratus 44
  vernus 76, 122

Lavandula 22, 44, 64, 99
L. angustifolia 99
Lavender 22, 44, 59, 64, 68, 99
Leaf-cutter bees 11, 12, 17, 44, 46, 53
solitary mason bee 9, 44, 50, 53, 105, 124
Leonurus cardiaca 22, 24
Leopard's bane 111
Leucojum vernum 115
Lithospermum
  purpurocaeruleum 85
Lonicera
  periclymenum 65, 73
  xylosteum 123
Lungwort 28, 32, 62
Luzula nivea 79
Lysimachia vulgaris 124

Macropis 124
Malus 48, 63, 88
Marjoram 25, 40, 82, 89
Mason bee 120, 9, 44, 50, 53, 105, 124
Meconopsis cambrica 83
Megachile
  ericetorum 44
  nigriventris 44
  willughbiella 17
Melitta haemorrhoidalis 9, 124
Melittis melissophyllum 76
Mignonette 42, 80, 124
Mining bees 53, 94, 101, 110, 113, 119, 120
Mint, Korean 67
Monarda 75, 90
Monk's hood 10, 65, 70
Motherwort 24
Mountain fleece 38, 90, 96
Mourning bees, common 94
Myosotis sylvatica 120

Nepeta × *faassenii* 77
Obedient plant 74
Onion, ornamental 77, 114
Ononis spinosa 22, 44
Orchard bee, European 48
Origanum vulgare 25, 89
Orpine 92

Osmia
 adunca 9, 78, 124
 bicornis 48
 brevicornis 124
 cornuta 48
 florisomnis 11
 rapunculi 9, 50, 53, 124
 truncorum 54

Papaver
 orientale 97
 rhoeas 64, 97
Perennial cornflower 9, 40, 62, 72, 97
Perovskia atriplicifolia 68
Phacelia tanacetifolia 64, 100
Phlomis 69
Physostegia virginiana 74
Plume thistle *Atropurpureum* 122
Polygonatum multiflorum 30, 79, 122
Polygonum amplexicaule 38, 90, 96
Poppy 10
 common 64, 80, 97
Pulmonaria 28, 32, 62
Purple gromwell 85

Reseda 42, 124
Restharrow, spiny 22, 44
Ribes alpinum 20, 21
Rosa 40
 multiflora 102
 rugosa 81
 spinosissima 101
Rose 10
 field 124
 many-flowered 102
 Scots 101
Russian sage 59, 68
Sage 9, 10, 35
 common 22, 24, 59, 68, 82
 meadow 11, 25, 122
 Russian 59, 68
 sticky 122
 whorled 122
 woodland 122
Salad burnet 25

Salix 20, 40, 48
 triandra Semperflorens 107
Salvia
 glutinosa 122
 nemorosa 122
 officinalis 22, 24, 82
 pratensis 25, 122
Scabious bee, large bryony 124
 common 34, 54
 grey-backed 124
Scilla 16
Sclarea 24, 56, 59
 verticillata 122
Sedum 35, 62, 92
Silene coronaria 78
Silphium 10, 84, 90
Snowdrop 117
Snowflake 115
Solitary mason bee 9, 50, 53, 105, 124
Solomon's seal 30, 79, 111, 122
Spiny restharrow 22, 44
Spotted bee balm 75
Squill, Siberian 48
Stachys
 byzantina 22, 27, 44, 56
 germanica 22, 25
 grandiflora 124
 officinalis 22, 26
 recta 22
 sylvatica 121
Stonecrop 35, 62, 92
Sweat bees 40, 100, 105, 119
Symphytum
 asperum 28
 azureum 122
 grandiflorum 28, 30
 officinale 28, 65, 122
Tanacetum 34, 37, 54
Tansy 9, 34, 37, 54
 purple 64, 100
Teasel, wild 12, 62, 84, 95, 96
Tetradium daniellii 103
Toadflax, purple 22
Trachystemon orientalis 29, 33, 62
Tutsan 123

Valerian 35, 68
Veronicastrum virginicum 74, 90
Viper's bugloss 9, 12, 78, 80, 124

Wallflower 28, 35, 94
Weld 42
Welsh poppy 83
Wild garlic 110, 111
Willow 10, 20, 40, 48
 almond 107
 goat 125, 124
Wisteria 56, 58, 124
Wool carder bees 22, 24, 26, 27, 59, 71
Woundwort
 downy 22
 hedge 121

Xylocopa violacea 47, 56

Yarrow, common 34
Yellow-faced bee 42, 53, 105
 common 42
 onion 12, 42, 124

Also available from Haynes

Bee Hotel

This book contains easy-to-follow instructions on making every kind of bug hotel, from the unbelievably simple stop-over hostel, to quite elaborate and ornamental luxury hotels. Start small and scatter your garden with drill holes in the right kind of wood, bamboo or reed tunnels in the correct places, and make screw holes available in masonry or garden furniture. Move on to more ambitious structures, like a planter filled with mud or shelving battons screwed into a wall, and then try one of the decorative hotels that will help to make your backyard start buzzing.

- 30 step-by-step DIY insect home projects
- Simple to make and easy to install
- Illustrated throughout with diagrams and photos

FROM ALL GOOD BOOKSHOPS • ISBN: 978 1 78521 658 9 • RRP £10.99

Bee Manual

The *Bee Manual* provides a complete and easy-to-follow reference to the intriguing world of the honey bee and the addictive craft of beekeeping. Aimed at the novice but also containing plenty to interest the experienced beekeeper, the *Bee Manual* presents no-nonsense advice, facts and step-by-step sequences, as well as plenty of relevant photographs and diagrams. Find out how to work with these amazing insects to enable them to thrive, pollinate, and produce a honey crop – and play a part in reversing the decline in the number of bee colonies.

- Everything you need to know about the honey bee, its life cycle and activities
- Choosing equipment and obtaining a colony
- Advice on swarm prevention and control
- How to harvest, store and use honey

FROM ALL GOOD BOOKSHOPS • ISBN: 978 0 85733 809 9 • RRP £22.99